Edited by

SAMUELE SANGALLI

Beyond the Limits

Consequences of Technological Revolution in Society

G&BP
GREGORIAN & BIBLICAL PRESS

FONDAZIONE
SINDERESI
praticare l'etica

ACTON INSTITUTE

Konrad
Adenauer
Stiftung

For the Konrad-Adenauer-Stiftung the exchange of opinions and open discussions are important instruments of civic and political education. The content of this book is not reflecting the positions of the Konrad-Adenauer-Stiftung, but it is based on the exchange of young leaders, which the Foundation aims to support and promote in their development.

Cover design: Serena Aureli

Layout: Yattagraf srls

Gregorian & Biblical Press
Piazza della Pilotta, 35 - 00187 Roma, Italy
www.gbpress.org - books@biblicum.com

ISBN: 978-88-7839-**372**-1

Table of contents

In loving memory of
Cormac Card. Murphy O'Connor
Master of overcoming challenges "beyond the limit"
through his humble and transparent manner
and sense of humour.

Foreword

The *Sinderesi School* has dedicated its annual research (2016-2017) to investigating "*Beyond the limits. Consequences of technological revolution in society*". In this volume, together with three scholarly presentations from various areas of expertise, the reader will find the research and the results of the lively debates among the young people who make up *Sinderesi School.* The unique methodology of this programme is the multi-year experience of the students, supported by qualified experts, who prepare and then discuss, in various workshops, the findings of their research. Here, the reader will find the fruits of their labour offered now to the broader public.
It is truly fascinating to become part of the intellectual journey of these new generations, as we accompany them in their lectures and through their individual study, research groups and monthly brainstorming assemblies. We have the opportunity to see the results of their education as the young undergraduates or graduate students deal, in a truly perceptive way, with contemporary social challenges. In this year's study, we see them engage with the main consequences of technological improvements and the huge changes they bring to bear on humanity.
As Msgr. Sangalli writes in his introduction, the essays gathered here offer: "interesting and inspiring points of view on the ethical challenges resulting from innovations regarding intellectual property, the cyborg project, neuroscience's data, the questionable beginning of human life, democracy in a 'post truth and fake news' era, the changes in the financial system and, finally, the discoveries for a new cosmology".
The Preparatory Document "*Young people, the Faith and Vocational Discernment*" of the XV Ordinary General assembly of the Synod

of the Roman Catholic Bishops asks: "What assistance is offered to young people to look to the future with confidence and hope, beginning with the richness of the Christian roots of Europe?" These contributions, from the young people of *Sinderesi School*, are offered to all who want to relate to the contemporary world in a competent manner. These findings demonstrate how through supporting each other and rejecting individualism, these young people are able to make a serious and thoughtful discernment, inspired by our Christian faith. Moreover, this book also demonstrates that the reflections and responses of young people, who will bear the responsibility of the future, are making a singular contribution to the public debate.
I praise this initiative of *School Sinderesi* and its methodology: a model not only for Italy, but also for all of Europe and beyond. It could truly represent one of the answers to the next Synod's call.

+ Vincent Card. Nichols
Archbishop of Westminster

Preface

This fine book presents the consequences of technological revolution thought by participants of "Scuola sinderesi" under the guidance of competent and experienced university professors. As I went through the book, I discovered that the contributors of volume attempted successfully to respond in one way or another to the following question: what means to be a human being or to leave in in a world constantly shaped and challenged by technology? From the study of genetics and intellectual property to that of cosmology, neurosciences, human structure, democratic procedure, bioethics and finance system as it relates to techonological revolution, the paramount question tends to be the future of human species and the type of values that contemporary society will pass on to the next generation. With a faith consciousness, the contributors of this volume repond to the anthropological question that we all face in cultures driven by the technocratic imperative. Using the words that Monsignor Sangalli borrows from the italian philosopher Umberto Galimberti, the contributors of this book grappled with challenging issues with a critical mindset showing how "in the human evolutionary process technology has gone from being a means, to becoming an end itself, the real subject of history. Not only has the potential side effects of our technology gotten out of hand, but the ability of man and the world to continue to exist is questionable." However, adopting a posture of laments and negativity will not help anyone in today's context in which technology not only determines how we inhabit the world but also who we are. We cannot help but reckon with the irrefutable truth that technology is not only needed but also unavoidable. The central challenge that we all have to face is then

to know how we can use it for our good and for the upbuilding of the world. Even though technogical progress raises important questions, how can we turn into an opportunity. In the era of triumphant neoliberal capitalism, we are challenged to turn technological revolution into a non-destructive and beneficial opportunity to contemporary societies? Sangalli argues that, since the ego tends to be at the center of our technocratic culture, education, which helps overcoming egoism, is the way to go.
I sincerely recommend this book for the dreams of the youthfulness, joyous approach to faith, stubborn passion to communicate and faithfulness to the Church that inform the contributors' reponses to complex questions pertaining to technological revolution. The authors of this volume are not the first to raise important ethical and anthropological questions with which contemporary men and women are confronted in this age of technological revolution. Nevertheless, what makes this book worth reading is that contributors are young christians who endeavoured to share with the rest of us the type of world they would like to live in and the set of values that should make it livable and enjoyable. Why don't we open up their dreams, hopes and values to see how far they may challenge us to fight for a better world and a life-enhancing technological revolution.

P. Jacquineau Azetsop SJ,
Dean of the Faculty of Social Sciences
Pontifical Gregorian University

Introduction

As I said in my introductory lecture at the beginning of the course: «The present edition of *Sinderesi School* stands in close continuity with the previous one,[1] which considered the challenge of "speed" as a specific qualification of this epochal change in human history. The factor that most of all has favored this flux is technology that, in its pervasive influence, has reached and transformed every aspect of the world» (Lecture n. 1).

There have been huge changes, undoubtedly, regarding communications, culture and political participation. On the one hand, technology has made possible new types of transportation that favors a growing process of inter-ethnic mixing; on the other hand, the prevalence of virtual reality has reshaped the economic and financial spheres. Medical technology has shaped humanity in regards to sexuality, to procreation and to the end of human life. If we look at technology's impact on religious practice, it has to be noted that online religious activity is becoming a relevant form of religious experience. For these reasons, my analysis seeks to explore and to get to the core of what technical evolution means in relation to the development of man and the world. According to Martin Heidegger, technology's domination represents the inevitable destiny of Western metaphysics, which is marked by nihilism. The famous Italian philosopher Emanuele Severino calls this phenomenon *technocracy*, a system that needs to increase its own power indefinitely. Another philosopher, Umberto Galimberti, states that, in the human evolutionary process technology has gone from being a means, to becoming an end itself, the real

[1] SANGALLI S. (edited by), *A Changing Humanity. Fast-paced living as a new model of being*, G&BPress, Rome, 2016.

subject of history. Not only has the potential side effects of our technology gotten out of hand, but the ability of man and the world to continue to exist is questionable. Despite these changes, the humanity of our so-called *technical age* has been marked by dissatisfaction that originates from various forms of depersonalization, such as dependence on computer media, depressive states, impossible work obligations, and different forms of ideological or religious fanaticism. We either interpret this epoch as an opportunity to take personal responsibility for the planet, or with the present trend we will abdicate each conscience to *technocracy*, which inevitably means a surrender to intolerance. In fact, we have learned from history that a similar ideology was at the root of the most disastrous experience and of the most heinous crimes against individuals and communities. Only a correct education of each conscience, through listening and by pursuing the goal of human "self fulfilment" in a respectful and responsible way, can counteract such a dark possibility. If humans are relational beings, their duty in the present time is to take care of the deep need for love without new devices or blackmail, dedicating more efforts to formation in discernment, to the education in affective maturity, to the cultivation of the "interior life" and to an authentic commitment to the common good. In this age of technology, we either slide into inauthenticity by taking advantage of these unprecedented technical possibilities, or we choose responsibly to submit ourselves to the discipline of a long and liberating journey of purification from our egoism in order to attain mutual self-fulfillment.

The education of "the self" should be originally applied to the evaluation of how technology could interfere with biology in the formation of human life. Dr. Roman Cholij from the University of Cambridge (UK) offered a presentation on how to deal with intellectual property, particularly regarding genetics. Following him, Professor Paolo Benanti from the Department of Theology at the Gregorian University analyzed what cyborg and cyborgization means to science and society.

Since many of the new technologies and their applications raise serious bioethical issues when applied to genetic resources and to human beings, Cholij takes a Christian ethical view of some specific areas regarding intellectual property protection through the use of patents. Many of the genetic resources possessing significant economic and social value are based in less developed countries, which are inhabited by native communities. To use this resource outside of a right patenting system and without benefits sharing, would be considered a misappropriation and even "bio piracy". Due to the powerful forces and competing interests involved, the relationships from genetic resources and traditional knowledge to intellectual property protection, this has been one of the most complex and controversial issues of multilateral diplomatic deliberations during the past decade. The goal of these good efforts has been to formulate an international legal instrument on the patenting of genetic resources consistent with the objectives of the Convention on Biological Diversity (1992). In this ongoing discussion, based on key principles of Catholic Social teaching, the Holy See has underlined the social function of intellectual property and the preferential option for the poor towards the common good, the end of all human law. Control mechanisms should accompany the logic of the market with the ability to patent a living organism, such as a plant, an animal or something genetically altered. So as to avoid inappropriate and incorrect language in the theological and ethical discourses regarding this issue, it is important to understand that a patent does not confer ownership on a private individual or on a company for something that is considered part of the common heritage of human beings. Rather, patents guarantee only a temporary legal protection, usually up to twenty years, against attempts by other parties to commercialize the patent holder's discovery or invention. Excluding the use of information and of certain techniques is very different from owning a piece of living property. Certainly, knowledge about a particular gene may have specific healthcare implications. For these reasons, public organizations involved in the Human Genome Projects mutually agreed not to patent their findings, but to immediately release any information into the public domain. The Catholic Magisterium likewise stressed this point, saying that discovery about the human genome should not become the property of a small group through patenting, but should be made available to the scientific community. According to Dr. Cholji, when intellectual property and patents are done in a way that respect the integrity and dignity of each person and in a socially responsible way, such patents would have an important role in promoting the work of genetic research and development.

Paolo Benanti, making use of the word *cyborg*, which was coined in the sixties, at first describes the five possible articulations of the interface between human and machine. The last articulation, representing an evolution that will change human nature, consists of a transgenerational fusion of humanity with technology that will allow the essence of consciousness to survive removed from the body, but with functionality, highly evolved feelings and a more developed empathy. It could bring about post-humanism: a universe that consists principally of information, where information patterns are more essential to the nature of being than any material biological substrate. In this way, humanity could eventually achieve immortality. In fact, it could become only a matter of technical choice of individuals and societies.

Passing from what was called anthropo-ontology to the new cyborg-ontology, these new beings, cyborgs, would not be bounded to the previous steps of evolution and could easily lose their "humanity" in a system qualified by complexity and emergence. By converging the areas of nanotechnology, biotechnology, information technology and cognitive science, the United States National Science Foundation risks achieving machine-based and dependent human enhancement. Cyborg will become an enhanced and complex human, linked to a combination of hard and soft interface technology, which would enable and augment information, acquisition, integration and cognitive capacity. Benanti asserts that the process of the cyborg has begun, however, "its trajectory and destination rests in our hands".

These three lectures provided the tools for all the members of our School to conduct their research. In this book we have collected into five articles the results of individual studies, research groups and monthly brainstorming assemblies, which describe how technology shapes the world in various areas. Two articles are deeply related to the "identity" of human life, as stated before. The first research group, led by Antonella Piccinin and Rafael Pou, L.C., under the supervision of Professor Franco Imoda, S.J., from the Gregorian University, offers the results of their investigation in a contribution called "Neuroscience and the Search for the Person": They take criminal law and human structure for their case study. We live in the so-called "century of the brain", and neuroscience is addressing the fundamental question of human identity, promising to reveal the underpinnings of our individuality. Neuroscience is a vast and hybrid discipline that investigates the human nervous system to understand humans and their actions. Its aim is to explain how an individual interacts with the organ of the brain or, if he is only just a part of it. The article moves from an analysis of the theoretical and practical implications of neuroscience in the law, to describe the most relevant philosophical and anthropological consequences on modern societies. According to certain results from neuroscience, could the principle of subjectivity be nullified? Could we recognize the capacity of discernment in human being is irreducible to only a sort of "software"? The history of philosophy has developed a dualistic approach (body-soul or brain/mind-soul) or a monistic (materialistic) one regarding the essence of the human being. Both of them appear to be a type of reductionism. We still do not know how a physical mechanism can hold qualitative states and generate an individual *self* by the simple fact of realizing certain functional states. Certainly, a theory that does not take into account the complexity and totality of the individual seems deeply incorrect. Both of these reductionist views appear insufficient to explain the constant interrelationship between the brain and body. The writers invite us to consider a third option, a holistic conception that understands "the man as person, conceived as a unified totality of body, mind and spirit/soul, whose corporeity represents the privileged place of its realization". Through an inter- and multi-disciplinary approach, we can take into account important human phenomena, such as free will and the fact that every human act is a single psychosomatic event with a formal (psychic) dimension and a material (neurological) dimension deeply interconnected. In order to evaluate the effective contribution of neuroscience in the investigation of the human person, it is important not to deny the multifaceted nature of the human being. In a wide horizon of research, science plays a fundamental role, but always in relation to other disciplines. A personalistic approach, which is more inclined to recognize the deep complexity of humans as holistic and dynamic beings, ensures the recognition of his intrinsic dignity. As the French philosopher Merleau-Ponty wrote: "in order to understand the human being the brain is a necessary but not sufficient condition".

The research group led by Giacomo Alfiero and Xavier Gutierrez, L.C., under the supervision of Prof. Ramon Lucas Lucas, L.C., from the Gregorian University, specifically focused on the ability of technology to create human life. Their contribution regarded "an alternative Bioethics", with the question on "when human life begins" for their case study. Combining a physiological approach and a philosophical one, they offer an anthropological reflection concerning the beginning of life, with an ethical consideration of assisted reproduction. Out of this conundrum comes an ability to inform our moral evaluation. It is a fact that, at the very inception of the embryo, embryonic genome activation (EGA) demonstrates the embryo's ability to activate gene expression to continue to foster its own growth and organization. At the present level of technology, embryologists have the ability to understand the capacity to which the embryo can act autonomously and in a highly organized fashion. Being so autonomous by nature, with an ability to organize and control various aspects of its own development for up to 12 days after fertilization, independent of maternal influence. Based on this, would not the human embryo deserve the same respect as every human being? Certainly almost every bioethical debate touches on the concept of personhood and "how and when" a concrete individual human nature could be recognized as "a person". Through the Universal Declaration on Bioethics and Human Rights, UNESCO hoped to resolve most of the bioethical issues and debates. Nevertheless, modern bioethics has become a major area of contention because of the controversial definition of the person. Many have followed John Lock's reductionist theory that locates the human person solely within the consciousness, or the need to be aware of its own existence in order to have a human identity. However, if personhood can only exist if an individual is conscious of its existence, how can we exclude from humanity persons in a persistent vegetative state or persons profoundly disabled? In this regard, the authors challenge the readers by asserting that it is hard to deny that human life unequivocally begins at conception. Therefore, before such new technologies concerning the beginnings of human life, modern societies should no longer delay a common agreement regarding the moment when the embryo begins to exist as a person, so as to weigh the moral acceptability of various interventions. The use and role of the embryo has been very controversial in jurisprudence, and a common final criterion has yet to be discovered. The article concludes by stressing the fact that if genetic impression occurs at the embryonic stage, as it is in all the other passages of an individual life, so we should surmise that there are no reasons to deny personal identity as a human being to this new individual. Acknowledging that the embryo is a human being from the moment of conception, the moral duty to legally protect it must be recognized to avoid manipulation and experimentation unworthy of his dignity.

Two other groups expanded the research to other aspects of society. They were engaged with impact technology's effect on politics, economics and finance. Nino Marzullo, Nicola Pigna and James Cleary, L.C., under the supervision of Professor Giovanni Moro from the Gregorian University, offer a paper on "Technology and Democratic Procedure", with a case study on how technology shapes democracy in social networks and in the "post-truth" era. Technological revolutions have transformed the decision-making processes of citizens, who are now more involved with personal causes, specific movements or spontaneous gatherings that spring up and develop from the internet. New technologies of communication are so decisive, that the preferences of the participants can change in the course of the interaction. A serious analysis on the use of the internet shows how public opinion is increasingly dependent upon the web, and misled, or at least influenced negatively, by fake news spread by websites lacking checks. During the campaign of the recent UK referendum, "Brexit", fake news played a decisive role in the final victory of the "Leave" party. We are in the, so-called, "post-truth era", when objective facts are less influential in shaping public opinions that the appeal to emotions and personal beliefs. Today, politics is played out on the internet and, especially, in social networks. The quantitative and qualitative analysis of three recent referendum (the abovementioned one in Great Britain, the proposed constitutional reform in Italy and the Columbian plebiscite concerning a peace treaty between the State and the FARC) represent the case study for this article. Social networks in Italy were able to foresee the resulting votes from public opinion and it offered at the same time a new survey instrument. In Britain research demonstrated how the progress made by those in favor of "Leave" was largely due to their communications strategy in the months leading up to the referendum. A similar surprise was found with the results of Columbian referendum where the government was unable to use social networks to convey institutional messages and to counteract fake news. The researchers' contribution clearly shows that, due to the greater immediacy it offers by conveying information, internet technology helps to bring about the "dematerialization" of traditional physical places for politics, and in doing so changes the organizations of our society.

Mariangela Lancillota, Irene Buzzi and Adrian Lawrence, L.C., studied the changes in the financial system, by focusing on impact investing as their case study. After a brief analysis of the evolution of economic theories, their contribution focused on the so called "impact investing" as an alternative financial market tool suitable to the new generations who are more aware of the threats the present economic system represents for the future of our planet. The spread of financial deregulation to most of the world markets needs to be addressed more seriously if we want to come out of this unprecedented global recession. Policy makers should combine macroeconomic policies with market-regulation measures. The principle of capitalistic accumulation, or continuous unlimited growth, is not only erroneous and unsustainable in itself, but it proves to be destructive in the end due to its environmental costs.

What is needed in the world is a new approach to economics, one that chooses a value that is not limited to economic and quantitative aspects, but to a broader social value, which will ensure that businesses always act fairly and ethically. In recent years we can note several alternative approaches such as the economics of happiness, the theory of decline, the economy of the gift and the advent of Corporate Social Responsibility (CSR). The latter teaches that corporations have responsibilities to society which extend beyond economic and legal obligations. The new generations, having witnessing the global recession from 2008, are no longer interested only in forms of philanthropy or classical alternative finance, but are willing to deal with broader financial concepts like impact investing. This one could be qualified as the ascent of "philanthrocapitalism" to answer to the needs of both the market and society. As the Rockefeller Foundations stated, impact investing is like "a set of investments in society, organizations, and funds with the intention of generating, in addition to financial returns, a quantifiable impact at both a social and environmental level". It is important to realize how new technology contributes to this, through the use of crowdfunding, bitcoins or blockchain, to the development of impact investing. The authors conclude their paper by underlining how, through recent initiatives, even the Holy See has shown interest in impact investing, since it is a system that in our time of profound changes brought on by the technological revolution can be compatible with the values proposed by the Social Doctrine of the Church.

The final contribution, "Cosmology – after quantum physics: shaping a new cosmology; with the case study on cosmic inflation and fine tuning", is offered by the group led by Lucia Cosmano, Marilena Pisani and Michael Baggot, L.C., under the supervision of Dr. Umberto Cannella and Professor Emilia Palladino. Their aim was to look at how new technology has changed our understanding of the universe. They have chosen to focus on a few representative theories of physical cosmology (the Big Bang theory, the inflationary paradigm, the apparent fine-tuning of the universe and the proposition of a multiverse to explain it) as the fruits of scientific and technological development. These theories have profoundly impacted how humanity interprets the cosmos. On the one hand, the Big Bang evolution and the Guth's inflationary paradigm seem to indicate a simple universe. On the other hand, the most reliable cosmological history available including more than 20 fundamental constants of physics, points to a cosmos finely tuned for our life, asserting what is called the "anthropic principle" of intelligent design. The challenge remains to interpret properly the scientific findings by themselves and in relation to other domains of human knowledge. Following contemporary philosopher William Lane Craig, the authors underline possible hermeneutics for understanding how technical advances in the vision of the cosmos might affect our anthropological conceptions. Is our universe a physical necessity or the fruit of pure chance? Using the Ockham's razor rule, many scientists seriously deny that the intricate network of life-permitting factors occurred in our universe by mere coincidence. The probable existence of an intelligence behind the cosmic design seems to transcend the spatio-temporal limitations of our cosmos and remains distinct, although compatible, with various theistic traditions. In human history, through various cosmological theories, science has tried to answer the question regarding the origin of the universe. The latest cosmological data, obtained through our advanced technology, still suggests an explanation for the fine-tuned features of our cosmos, an explanation not far from the meaning that humanity has always expected there to be.

The ethical challenges resulting from innovations regarding intellectual property, the cyborg project, neuroscience's data, the questionable beginning of human life, democracy in a "post truth and fake news" era, the changes in the financial system and, finally, the discoveries for a new cosmology represent some of the challenges of our technical age. In order to address them, what will be needed are educated people who are well aware of the risks and the gains. The participants of *Sinderesi School* have tried to become such educated and informed students through their research.

As has happened in recent years, *Sinderesi School* now publishes all of the contributions to share these debates on timely and relevant issues. The ideas collected here belong mainly to young people. It

is the fruit of their study and a demonstration of how it is possible to create active pockets of new thinkers able to offer interesting and inspiring points of view resulting from their willing to be present to the contemporary world in a competent manner.
Like previous ones, this edition is possible thanks to the invaluable contribution of all the above mentioned people, particularly to my tireless primary collaborators of the *Sinderesi School*: my Assistant, Dr. Antonella Piccinin, the heads of the five groups of research and the five Brothers from the Legionaries of Christ, whose community generously supported them. Profound gratitude is also due to the Pontifical Gregorian University, especially the new Dean of the Faculty of Social Science, Fr. Jaquineau Azetsop, S.J., the Directors of the "Alberto Hurtado" Centre and of the Gregorian University Press. Special thanks for the financial support of the Rome section of the "Konrad Adenauer Foundation" and the Rome affiliate of the "Acton Institute". I must also express my sincere gratitude to my colleagues at the Congregation for Bishops in the Roman Curia, Msgr. John Cihak and the Rev. Luke Sweeney, for their valuable support.

Mons. Samuele Sangalli

First part

Lectures

Lecture n. 1

The conversion of the ego as priority of the techno-age

Mons. Samuele Sangalli
Coordinator of Sinderesi School
Pontifical Gregorian University

1. The Age of Technology

The present edition of *Sinderesi School* stands in close continuity with the previous one (Sangalli 2016), which considered the challenge of "speed" as a specific qualification of this epochal change in human history. The factor that most of all has favored this flux is technology that, in its pervasive influence, has reached and transformed every aspect of the world. Our day is defined as "techno-age", in as much as technical supremacy is unquestioned. It also applies to man as defined by neuroscience, genetics and bioengineering, in terms of not only care and healing, but now redefined through his perfection and even the overcoming of that which, until now, has only been defined as man's "nature". The scientific debate profoundly questions what will become of that vision of the world that until now is known as "common sense".[1] It should be noted that the massive contribution of technology has radically changed the various "areas" or "spheres" in which human society lives and develops. Let us try to keep this in mind. First, this idea was made evident last year at *Sinderesi School*, in terms of *communication*: The internet is not a virtual world but is part of real life. Millennials (Czerski – *Millennial Manifesto*), or those who refer to themselves as the "cloud generation" (Facci –

1 «Che cosa rimarrà – ci si interroga alla luce della meccanica quantistica – degli oggetti fisici che popolano il mondo, i quali occupano regioni di spazio e tempo ben precise, possiedono caratteristiche definite e sono chiaramente separati e indipendenti l'uno dall'altro? E ancora: dovremo abbandonare – si domanda chi abbia studiato la teoria della relatività di Einstein – le nostre idee intuitive di tempo e divenire? E infine: ci sarà ancora spazio – ci si chiede alla luce del vertiginoso sviluppo delle scienze della mente e del cervello – per le persone, intese come entità capaci di esperienze coscienti, stati mentali intenzionali e agire deliberato?» (Lavazza 2016: 7).

Valorzi – Berti 2013), live to be "connected" but with incredibly poor relations. Those who have grown up living "online", conceive it as a free world that is constantly changing, a shared "extended" memory, that is able to supply any information deficit in social life, now structured as a horizontal *network*, without hierarchical criteria. As regards *social networks*, there is constant continuity between man and machine. It is clearly a community in continuous expansion, where *machine learning*, the automatic process of acquiring knowledge, guaranteed by an algorithm of universal learning, in the intentions of its promoters, will be able to bring to fulfillment the idea of an knowledge that is integral, synchronic and diachronic, and which is a guaranteed part of our everyday lives (Domingos 2016).
This is clearly in action also in the change of the paradigm by which we acquire and share knowledge, which is the *culture* of a people. Academia, dominated by *publish or perish*, cultivates the myth of the segmentation of scientific knowledge, which leads to specializations increasingly incommunicable beyond narrow circles. Thus, it creates new learned/technicians deprived of that *paidéia* that marked the real "masters" of the *Universitas Scientiarum*. Similarly, in the lower grades of education, through indiscriminate access to information, we are assisting instead the delegitimization of the figure of the teacher, and the return of mass illiteracy, primarily understood as a generic and superficial factual knowledge, indifferent to that necessary critical investigation to achieve the "*ritener l'aver inteso*" (*Paradiso*, vv. 41-42), without which there is no true knowledge. We have already discussed, in a previous edition (Nicotri - De Luca 2014: 91-122), to take note of how such a deprivation of the idea of culture could happen, subjected to the continuous and comprehensive flow of information, which by now no one is able to make any sense of it. This is likely to result in loss of education, arguably in a society that by now suffers from the lack of creativity of thought from many fronts and challenges, both political and social, of the contemporary world (Gimmelli - Tilotta 2014: 281-323).
Technology, as we have defined it in terms of information technology, could instead foster new democratic ways of *political* par-

ticipation, marked by the concept of active and digital citizenship. In this regard, I refer to what was discussed in the *Sinderesi School* on the relationship between democracy and citizenship in a global world. There it was asserted that: «E-democracy, open data, open government, in light of the development of digital technologies and the new dimensions of the time factor in the Age of Globalization, are terms that the world of politics will always feel closer» (Piccinin - De Napoli 2012: 210). Summarizing, obviously, the awareness that, on one hand the Internet can never replace the human mediation. On the other, the network is able: «to authorize encompassing exactly relationship as the fundamental element in the development and the formation of general and collective political thought, that looks not only at depoliticized individual and selfish interests» (Alfiero - Piccinin 2014: 113-114). In the Age of Technology, it is necessary to avoid that the political arena becomes a populism without responsible participation.
Indeed, the world is moving, for not only the great migrations currently taking place, but also because technology has made possible other types of *transport*, where they occur "without travel destinations". As I underlined last year (Sangalli 2016), a shift should be noted in the traditional forms of movement. There is greater fluidity of intentions and change of plans, reflecting the complexity of contemporary society, with a growing process of *inter-ethnic mixing* that tends to create widespread and strong bonds between communities living in different countries. In the end, mainly due to tourism or to moving for work and for study, there are various forms of contact between different populations. These are results of personal journeys, and – most of all - thanks to the computer network, spread over the last 20 years, allowing 41% of the world's population stable connection among themselves, without moving physically but only virtually, thus losing the human experience of traveling.
The paradigm of the prevalence of the virtual on reality, this singular expression of technology's domination, then takes place in the *economic* sphere, characterized by the imposition of finance, which has dethroned the centrality of work and caused the progressive increase in inequality. With the gradual elimination of the

restrictions on financial activity, investment banks transformed themselves from supporting productive activities and savings, to speculators looking to make artificial profits from financial engineering. Therefore, it became very much practice to make easy money from speculation, which has stopped the propensity to invest and risk in real activity, causing an inability to resume the recovery of the so-called "creative phase" of the economic cycle to the point where we speak of a *secular stagnation* weighted down by the so-called *jobless growth*. It means that when there is productive growth, it lacks the usual increase in jobs. As happened in the shift from the agricultural to the industrial era, so today, at least, in the most advanced countries, technology replaces man, who should therefore retrain with new criteria that is towards an occupation that is service oriented (Teulings - Baldwin 2014).
The will to power, particularly in its explicitly *performative* model in finance, ultimately is reflected also in the very arrangement of humanity, conferred to medical technology, applied to *sexuality*, to *procreation* and to the end of life or to *death*. The anthropological shift, favored by the technological revolution, comes down to the processes of redefinition of the events structuring the conditions of opportunities in people's lives. These are seen again through a constructivist logic with the project / the imperative to reshape reality according to the wants of the subject (Sandel 2014), without normative obligations for human deliberation based upon the previous bond of ethics / norms. This is done as well by passing through a radical reinterpretation of human sexuality, as it intersects with various emancipatory and libertarian instances. The tensions, arising from the achievements of biomedical technologies, record all the work of a world in which science goes faster than the ethical understanding. Even those who, in the name of the new liberal principles, look favorably on the slippery slope from moral absolutes of the framework of natural law cannot fail to see the danger of the "technocratic" drive, or that "the collapse of the traditional framework does not prevent others from seizing the context, at least in part, of these areas. They influence us by shaping the meaning of these areas of life and, therefore, the moral possibilities of thought, experience and action" (Donatelli 2012: 112).

If we look at the world of *religious practice*, it is only right to note what is being impressed, primarily on the new generations, by the presence of the computer, which raises a number of questions. As pointed out by Antonella Piccinin's recent research: «Religion and the Internet seem to have become so intimately and integrally intertwined that there are those who are now suggesting, paradoxically, that online religious activity could become the dominant form of religion and religious experience in the next century. .... This issue raises a number of questions: Is the enormous variety of information shared on the web - and more specifically religious information - truly worthy of a proper expression of the freedom of religion? Or else, is this fundamental right perhaps something more, or something else than what appears on the web? Is the freedom of religion guaranteed by the Internet? How is it possible to distinguish between liberty of expression and harmful / hateful speech, which, for obvious reasons, needs to be censored in the public interest?» (Piccinin 2013: 20-26).
Those described so far are some of the most important areas in which we can see the imposing effects of technology. It thus shapes itself as one of the principle products of cultural evolution, which qualifies the human difference from only biological evolution, which is proper to all the other living beings that inhabit the earth. Because by now the driving force of technical innovation has even arrived to inspire new scientific discoveries, and thus it virtually heavily influences the history of humankind, which way we can (or must) recognize it in relation to the development of man and of the world? (Sequeri 2015).

2. Destiny or responsibility?

Martin Heidegger is the starting point for a critical awareness of our theme (Ruggenini 1998: 225-264). In his philosophy of history, there is the intuition that technology's domination represents the inevitable destiny of Western metaphysics, marked by nihilism precisely because of that will of power in the concept of truth later adopted from Plato. That state of complete forgetfulness of being, denounced since the publication of *Sein und Zeit* in 1927, led to the submission of the foundation to speculative calculations and then to scientific reason, which is oriented in the end, through technological instruments, to manage powerfully the organization of reality with the illusion of negating finitude. The myth of the infinite production not only resolves, but even reshapes the needs of man, in an attempt to obliterate as well that which is proper to him.[2] Heidegger writes that: «in man's worldwide imperialism, organized technically, subjectivism reaches its highest peak. From here, however, it will be lowered to the level of the uniformly organized plan, to find their own arrangement» (Heidegger 1963: 102-103). Thus the history of the world would be resolved as a history of the decline of reason, which runs out in the technical dimension of its unconditional will of self-assertion on its natural energies. However, the leveling, the overall uniformity that such a perspective omnivorously produced, points out the insufficiency. Thus human existence is finally called upon to decide, or defending its own finitude by opening itself to the mystery of being which constitutes it, or resigning oneself to delegate to the calculation of technological reason the solution of all their problems. Western man can still - asked Heidegger dramatically in *Gelassenheit* in 1959 - not to be left exhausted from technological thinking, but to open himself up to a fuller reflection, not addressed to the sole use of things but also in search of their meaning? Unfortunately, the loss of such "deep thinking" and the establishment of technological thinking as the only truly valid thought form constitutes, to the existentialist philosopher, the true great threat to humanity, which is more dangerous than the Third World War. Such reductionism, in fact, would jeopardize the very essence of man, his ability to still be open to the mystery of being.

2 «Il soggetto non si avvede pertanto di aver ormai perduto tutto il potere su cui procedeva sicuro, nel momento in cui affida alla tecnica la responsabilità di risolvere i problemi che essa stessa produce: sovvertimento degli equilibri naturali, produttivismo sfrenato, il cui unico fine è alimentare consumi che domandino sempre nuovi prodotti e inesausta capacità di produzione, livellamento dell'essere dell'ente, riduzione dell'uomo stesso a materiale d'impiego, deresponsabilizzazione dell'esistenza di fronte alla suprema razionalità e provvidenza della tecnica, riduzione all'insensatezza di ogni domanda che non si subordini alla logica dell'assicurazione mediante verifica di compatibilità» (Ruggenini 1998: 246).

Emanuele Severino recovered with particular deepness the Heideggerian reflection on the theme we speak of as technocracy. According to Severino, this *apparatus* or power structure, today allied and an agent by way of capitalism, is effectively inhabited by the will to increase the infinite capacity to achieve goals. Thus it tries to soothe - or at least make more bearable - the anxiety of immortality which distresses man (Severino 2009: 116-125; Severino 2012: 199-201). Severino writes: «the term *technocracy* points to the prevalence, above every other form of the will, of the will to always more power. ... The individual wants to use the System to break free from pain (or rather from the way in which the pain occurs in the different renderings of it), and ends up subordinating this liberation to the infinite expansion of the System. But this is to realize that *every* need of man is a more or less disguised form of his *essential* need: the need to indefinitely increase his own power» (Severino 1989: 107; 117).
In this context, a detailed analysis, to describe the profound transformation of man in the Technological Age, is done by Umberto Galimberti in his *Psiche e techne* (Galimberti 1999). He first strongly argues as if we have not already taken account that, in the era of technology, the hermeneutical categories to describe the man and the world have essentially changed. In fact, technology is no longer the instrument, or neutral object available for our choosing, but: «it is our *environment*, where the ends and means, the goals and thoughts, conduct, actions and passions, even dreams and desires are technically articulated and have need of technology to express themselves» (Galimberti 1999: 34). In this sense, since technology was born as a remedy for survival due to the biological inadequacy of man, it should represent any expressed *essence*, through different forms in various cultures, the way in which humanity obtains its own adaptability and stability in the world. If, therefore, the presence of technology coincides with the human evolutionary process, this "quality leap" which has recently taken place, has made possible its increasing "quantity" for the realization of any end. Technology, according to Galimberti, from mean has become the end. Not because it proposes it in itself, but because only through it can humanity achieve all of its goals and

desired ends. So technology has become the real subject of history, putting nature into the background and man as its functionary. In fact, man realizes himself, to the extent that he becomes a servant to the Technological System with which – as Heidegger and Severino wrote – he is identified. Therefore, we participate in the redefinition of the same categories of man's self-understanding. The notion of individuality dies, because the actions of the subject are understood as calculated possibilities of the Technical System, which functionally provides and is described in the way of their execution. For the first time in history it is possible to "make society" without having human beings needing to be connected in bonds according to their personal nature. This begins a process of de-individualization and de-privatization, proper to societies that are homogeneous and conformist, where the crowd is no longer to be understood as the concentration of many, but rather as a collection of millions of individuals, everyone doing the same things but in a solitary way. The means of communication have fundamentally changed the way of experiencing reality, eliminating direct contact with the world - basing every communicative need - for a mediated representation, which configures all as a collective monologue. The conscience of each individual, deprived of its own uniqueness because of personal experience, thus tends to overlaps interiority and exteriority, profoundness and superficiality, activity and passivity. Human reason tends to be reduced only to a procedural instrument, the business gains as the sole criterion of truth, ideas reduced to mere working hypotheses, ethics as actualization of what can be done, and finally politics as a passive arrangement for the technological administration of man. Galimberti argues that the pervasiveness of technology has opened an interpretive space that has dismissed the horizon of its nature, in its fixed dimension and inviolability, and from history, where technology was only instrumental to man's dominion over nature. In fact, today human nature is genuinely changed, according to the unlimited changeability made constantly and progressively possible from technical development. For the first time the understanding of the possible effects of our product has gotten out of hand, leaving us "emotionally illiterate" in front of

the unstoppable spread of technical progress that now grows by itself, outside of every horizon of meaning. Taking all of the above lessons mentioned, Galimberti also warns of a dark *fate*: «if we are not able to ask ourselves about the heights of the general and global dimensions without lacunae of the technical work. Each of us will remain ensnared in that individual irresponsibility that will allow technological totalitarianism to proceed undisturbed» (Galimberti 1999: 48). He also denounces how this "nihilism of technology" would be very dangerous because it not only concerns the question of the being and of man, but it would be able to question the same capability of man and the world to continue to be. From here, he also proposes, through his works, an action of *psychic expansion* because - even if the technical predominance is now inevitable - it will avoid all that happens without awareness, condemning man to insignificance.
In that direction, of the active and reactive at the awareness of the epochal change produced by technology and of the degeneration into *technocracy*, it seems to me they are now movements of different political and social instances, including everything related to the care of the planet in ecological challenge due to environmental degradation. One for all: Even Pope Francis Encyclical *Laudato Si'* uses the term *technocratic paradigm* (*Laudato Si'*, nn. 106-114) to describe the "throwaway culture" that imposes globally as the only model, which is severely threatening the very survival of man and the planet.
However, given the alleged pervasiveness of *technocracy*, is it still possible to avoid it? Or is an inevitable fate that the previously mentioned philosophers are proposing for humanity as inevitable destiny - in which the West would be the forerunner - which flows into the nihilism of the *post-human* characterization by competitiveness without borders? (Vaccaro 2009).
In this regard, it must be noted that the man of the Technical Age would be marked by dissatisfaction. This does not derive so much from the absence of those forms of economic and social stability from their past. As pointed out by Pessina, the current dissatisfaction is: «rather one that derives from the fullness of a present saturated with possibilities, where anxiety for the future, of per-

formance, of possible and ultimate improvements in the quality of physical and mental life is transformed into a duty. This is a new categorical imperative that leads us to break the temporal and biological thresholds in order to appear, even before a new existential youth, a type of *indefinite future*» (Pessina 2016: 8). Such self-dissatisfaction is the product of this culture of immanence, which, thanks to technical equipment, wants to ensure the welfare of material life at any cost and forever, by subjecting man to the depersonalization of continuous adaptation to technological standards, in just the same moment in which there is an obsessive exaltation of subjectivity. In essence, the individual feels conquered, emptied of its individuality, from the same culture which in a Promethean way tends to project it towards an unchallenged spatial and temporal dominion. Such forms of depersonalization assume the traits of dependence on computer media, or of various depressive states, as well as impossibility of a rhythm of work, to the various forms of ideological or religious fanaticism. Can this discomfort of contemporary man truly be ignored? Such a humanity seems to disperse like the sand[3] in a future in which it will no longer be master.

Reducing humanity to *destiny* means that the specificity compared to the *determinism* that governs other forms of life in the biosphere is not recognized. Such a renunciation of personal responsibility masks a *choice*[4] of subjection of man to the "Power" alternative to the "Care" of God, which is an orientation to dominion as an end in itself which tends to "turn off" the light of the responsibility (the heart of awareness) through the climax of megalomania as the sole reason of existence.

Thomas Aquinas had identified the root of such *hybris* in the Devil's sin: «*...peccatum diaboli in hoc ponit quod est sua potestate delectatus ... Qui habere finalem beatitudinem per virtutem suae naturae,*

[3] «L'uomo di sabbia è l'uomo di un itinerario. È il testimone di una umanità che cerca un passaggio, che tenta di aprirsi un varco. È una figura terminale, ma anche qualcosa che sta emergendo. Sta tra due mondi, nel momento in cui una cultura si sostituisce a un'altra» (Ternynck 2011: 12).

[4] «L'io resiste ad ogni sua negazione perché sempre si affaccia, ospite inevitabile, alle soglie e ai confini dei modelli teorici e delle pratiche effettive che indicano mete di autorealizzazione, percorsi di felicità e di benessere» (Pessina 2016: 3).

non ex gratia alicuius superioris, sed proprium Deo, manifestum est quod quantum ad hoc appetiit diabulus Dei aequalitatem; et quantum ad hoc etiam appetit Deo non subici, ut scilicet eius gratia supra suae naturae virtutem indigeret. Et hoc etiam congruit ei ... quod diabulus non peccavit appetendo aliquod malum, se appetendo aliquod bonum, scilicet finalem beatitudinem non secundum ordinem debitum, id est non ut consequendam per gratiam Dei» (D'Aquino 2003: 524). A similar attitude of proud self-sufficiency with regard to the fulfillment of oneself and the world, which is a dark force that dominates the world with its enticements, is denounced by Solov'ëv, at the dawn of the contemporary age (Solov'ëv 1975), as reality that keeps in check the world: «Evil is a bad will, a hatred against God. ... God Himself, save depriving freedom, cannot force a bad will. The moral optimism ... is unable to measure the depth of the roots from which evil springs: all ends well» (Solov'ëv 1971: 122).

Benedict XVI, in that profound meditation about the human history that is the encyclical *Spe salvi* (Benedetto XVI, 2007), applies this *sin of the devil* (*peccatum diaboli*) to man himself, in this change of mentality that underlies the big turning point at the beginning of modernity, with the dawn of techno-age. Going back to Bacon's *Novum Organum*, Benedict XVI points out the cultural shift made in those decades in which, through the new correlation between *experiment and method*, man has been put in a position to arrive at an interpretation of nature in conformity with its laws – *juxta propria principia* – and thus to finally achieve the triumph of art over nature. This is a new and deep correlation between *science and praxis*, a synergy that makes authentic knowledge the power of man over nature. The aim is for the construction of a new world: the kingdom of man. The *restoration* or *redemption* of man does not wait any more from religious faith, but from the newly discovered link between hope and experimental science, relegating instead faith to a private affair (this is the process of secularization!), which in some way becomes irrelevant to the world. Thus was born a new type of hope: *that which is no longer Christian*. It turns into the concepts of *instrumental reason* and *freedom*, seen in opposition to the previous tradition. A similar project is fully realized in our postmodern tech-

no-age. Once again Benedict XVI in *Caritas in Veritate* stated the consequent risks: «Technological development can lead to the idea of the self sufficiency of technology itself when man, asking only the 'How', does not consider the many reasons which led to the act. It is for this reason technology can appear with an ambivalent face... The process of globalization could replace ideologies with technology ... This vision is now so dominant the technological mentality overlaps the true with the possible. But when the sole criterion of truth is efficiency and utility, development is automatically denied...» (*Caritas in Veritate*, n. 70).
In alternative, to those who are subjugated to the technocratic idol and reduce the meaning of human existence: "to the great epic of the species, begun in biological evolution and prehistory, and then entered into recorded history" (Wilson 2015: 147-148), the Pope stresses the urgency of an intelligence that, not evading its responsibility, is able to understand technology with new eyes (*Caritas in Veritate*, n. 77).
One of the lecturers in this session of *Sinderesi School*, Paolo Benanti, in his ponderous volume of reflections on this theme (Benanti 2012: 407-409) denounces as many thinkers do the uncritical assumption of Heideggarian formulation that would have automatically identified technology as a dehumanizing force, independently of the use that they face. Considering that all this would be just an *inevitable destiny*, they have forgotten the concept of conscious responsibility, which always ultimately adheres to the *peccatum diaboli*, placing their freedom in service of what is bad. Letting oneself be carried away by a materialistic and immanent interpretation of the result of neuroscience and biotechnology, they have consented to that reductionist reading of man. They did so omitting the recognition of the ontological value of the principle of non-contradiction – as we refer to from last year (Sangalli 2016) –, which holds almost inevitably that, through the reprogramming brought about by technical action, humanity will be re-invented.
We know from history that a similar ideology is at the root of the most disastrous experiences of the most heinous crimes against individuals and communities; obviously all justifiable in the name

of *the process or the system*, in this case that which is technocratic. Against similar prospects, there is the existential cry of the defenders of the sanctity and inviolability of each human person, irreducible to an evolutionary moment of humanity. Such resistance to the attempt to limit the self or the conscience to the results of the nervous system is based on the spiritual quality of the human being. Therefore, they make a distinction between meaning proceeding from formatted matter and the consistency that freely expresses itself and relates the personal and accessible spirit – as recalled by the phenomenologists (Stein 2013) – only through the experience of themselves.

After all, "to dissolve" the subject in *destiny* would also mean to surrender to intolerance. To demand instead the impossibility to renounce personal conscience as the original and stable principle, unique subject of all human sentiments and operations, it means to recognize certain "moral and therefore juridical absolutes", as the same life and liberty of persons, which are in place to safeguard the individual. Pessina writes about it: «Behaviors which violates values are hoped to be prevented: for example the dignity of workers (workers' rights), the freedom of movement of citizens, the destruction of cultural patrimony, of the natural environment, are not part of the order of intolerance, but of justice. Which brings us back to the fundamental question: only guaranteeing values that are not arbitrary can we avoid intolerance. In the absence of these, each can accuse the other of being violent and intolerant, making fanaticism the dominant form of relations. Other investigations are not needed to be able to say that if you abolish the reference to any truth, if you deny that we can affirm the truth, it is much more easy to espouse, like you would not believe, intolerance» (Pessina 2016: 202).

If this is true, rather than pursue, according to the technocratic project, the megalomania of *peccatum diaboli*, the man of *techno-age* is called to a responsible education of listening and of pursuing, what Thomas called the *finalem beatitudinem*. This concept that we, with more extensive language, would identify as the mystery of "self-fulfillment" to which, through every good path of wisdom or religion, every man is obliged to refer.

3. The Primacy of Experience and the School of Discernment

With obvious Christian language, Søren Kierkegaard identified such a path of truth as one in which "Placing oneself in relationship with his very self and wanting to be that self, the 'I' bases itself, transparently, in the Power that has placed him" (Kierkegaard 1981: 179). Such a humility and openness to question before the mystery of life is, in fact, the essence of the religious spirit, and also sociological investigation (Garelli 2016) indicates as present man's restless questioning of technology, despite the overcoming of traditional beliefs. This need requires a reaffirmation without hesitation, first of all of those "not arbitrary values" that we have mentioned: the care of the soul (Mancuso 2007); that *gnòti seautòn* which, in evoking Socrates, is at the beginning of the adventure of the critical thought to human wisdom.

It is written that not from the outside but from the heart of men comes evils and bad acts (Mt 15,18). Thinking then of such a path as the surest ways of seeking the truth, and, therefore, of the good cohabitation with technology, we are able to recover a profound agreement with what Pope Francis affirmed at the beginning of his pontificate about truth: "It gives of itself always and only as a journey and a life" (Francesco 2013). The Pope always recalled the purpose as: "The understanding of man changes with time, and so too the human consciousness deepens. We think of when slavery was accepted or when the death penalty was accepted without any problem. So one grows in the understanding of the truth" (Spadaro 2013: 475-476). It is an echo of *Gaudium et* Spes n. 16, the famous conciliar text on the primacy of the conscience. No this, of course, means indulging in epistemological and ethical relativism (Decosse 2014: 156-169), but rather, deepening the interpretation of what is meant, taking account that the *absolute* is a relationship. On the other hand, it reaffirms the primacy of being, not as a dynamic of power, but as event / revelation. It educates our thought to the continual understanding of an encounter that occurs gradually in history. This meeting questions the taking of responsibility and a slow understanding of the historical forms in which are given the links between men, because it avoids both the sterile demonization, and the superficial acceptance, and thus ignores the consequences of certain current and future lifestyles of the social structure.

If man is a relational being, his conversion in the Age of Technology, means, first, help him be on good terms with himself, as a reality marked by relations and the call to relations. Educating in conversion, in that great jumble that is the human heart, it is the necessary precondition for then facing competently and responsibly the complex challenges that presents themselves to us before technological development, on which - at least on some of the most significant! - This year we will focus in the research work of the various groups.

First recall that the primacy of *relationship*, as qualifying human identity, is realized as conversion to the "need for love" that everyone has. Much hyperactivity, amplified by the possibilities offered by technology, comes from escaping relationship: one is afraid of meeting people, of feeling responsible for people, of mentioning their own weaknesses and of becoming dependent upon each other. Another person, in a true dialogue, can always make us feel insecure, because we do not know his reactions; we do not know his needs. Therefore, we are afraid to bind ourselves to them, not knowing where the other wants to take us. We are afraid of this relationship, we are afraid of the other's weakness, and we are afraid of our own weakness - we do not want to look at it (Vanier 2015). So we flee before it in activity, in work, in the search for purely material values, and, little by little, it hardens us, eventually we come to identify ourselves with the same technological instrumentation that we use.

The spiritual tradition, present in various religious-wisdom experiences of humanity, has identified conversion, a condition for real relationships, as an *education in listening* (Špidlík 1997). For not ignoring its own responsibilities, it is necessary we become true listeners by being able to be with our wounds, our vulnerabilities, our fears, our weakness, with that extraordinary mixture of weakness and strength, ignorance and knowledge, love and hate, tenderness and toughness that is who we are. We have to accept being only small beings in an immense humanity, looking for af-

fection, safekeeping, esteem, fulfillment, and love. To have much fear of failure, of being trampled, of being insulted, rejected and of offending and dismissing. In fact, those who have the courage to be with one's own frailty (Borgna 2016: 23), one's own weakness, limitations, discovers in himself the power of the deep sense of solidarity with others, of basically identical poor creatures in need of affection as I do, and the "original" invitation, of the child created and not someone who creates themselves, trust in other people's presence, in the good intentions of the other and in hope towards the good, clear, objective, which is love and acceptance, respect and appreciation of everyone.
Such intentionality comes together in the discernment and in the education of feelings; a spiritual practice now mostly forgotten or relegated to those who choose particular forms of consecrated life. Yet, we recall last year (Sangalli 2016), the "golden rule", as the imperative principle of the "good life", as the criterion of life for all, not just for "consecrated men and women" (During 2015).
For walking according to the "golden rule", the only safeguard for humanity yesterday and today, the education of one's consciousness is required. The others, in fact, reveal us to ourselves. Undoubtedly, for the man of technology, different is the socio-cultural context in which this takes place "in the conversion of the ego". However, it is the same "introspective" art, of "discernment of spirits" (1Cor 12,10; Rm 2,18) which moves within us, because he would be able to persevere in what leads to the good and to "neutralize" what objectively produces destruction and death (Ignatius of Loyola 2016). The discovery and cultivation of the "interior life" arises today, in a way that can no longer be postponed, as a condition of the possibility of humanity's survival called to exercise its own responsibilities in a context of "technological power" without precedent.
At the school of the great spiritual masters, it should be recognized that man lives constantly in the midst of changing passions, thoughts and desires. This is a variety of psychological states which, because he does not endure them passively in a state of subjection, requires a moral commitment that they "understand and order" according to universal criteria summed up in the golden rule.

Today, mostly, this orientation is accomplished basically through the often unacknowledged logic of "I can feel it, therefore it is good", ignoring how such a criterion really frequently hides the ego selfishly guiding the individual, that is introduced so surreptitiously, it paralyzes the person with respect to every good way. It can only be achieved through a patient detoxication from the same "Ego-Imperious" (Ameli 2014), which is a bitter medicine taken for healing.

Therefore, there *should* be above all a "*vigilance*" concerning "I feel", noting that the previously mentioned Delphic oracle was admonishing us to be very careful with respect to the activity of our interior world. There, in fact, negative thoughts insinuate themselves according to a progression that the Fathers identified as a stew of contact, conversation, combat, consent and passion (Evagrius Ponticus 2008; Cassiano 2007). It is not possible to prevent the entry of passions into our souls, but rather to confront them in a long process of identification and *transformation*; that which have their origins in the various doctrines about the virtues and the vices. Referring to the lessons of Thomas Aquinas, who exemplifies how elaborate Western spirituality is, I recalled that: «Man, according to Christian wisdom tradition, is able to *transform* the passions: from *threat* to the balance of a good and sound soul. Thomas, holding the vicious passions as genuine *opportunities for growth*, speaks of pride as a stimulus for excellence in the good for love, envy as an invitation to zeal, a righteous anger as passion for justice, sloth as strengthening for resistance, avarice as the engine of hard work and the entry into service of shared responsibility. Basically, we would say, very *pragmatically*, ethics, for Thomas, represents the crucial battle for truth and justice conducted by human reason. It is a struggle that is able to bring out man's best energies, to be used for the realization of oneself and therefore the building of society» (Sangalli 2015: 95-102).

Obviously, there are not only negative emotions, but also positive inner movements, of generosity, patience, dedication, acts that will lead the man to a virtuous life. Since then, in the end the "*moral stature*" of a person is the result of the habits to which he has applied his own freedom. A deeper awareness of the passions is necessary, which tend to shape the determination of man's free will in his daily actions. To ignore the influence in us of sadness towards present evils, the fear of the future, the pleasure in enjoying present goods and concupiscence towards the same expectations in the future, it means not to wrestle with the peace of soul, which – we repeat – can only be the fruit of a diligent attention to these impulses that demand to be "*ordered*" (John Climacus 2005).

However, the administration of the same is not possible if you do not also learn to take into account the diversity of temperaments, or natural reactions, which qualify the various subjects impacting with such external influences. In fact, an impulse generally is a prompt reaction to a stimulus, but it lasts only for a short time. A choleric person will act lively and in an enduring way. A melancholic person responds slowly, but for a long time; unlike a phlegmatic who is slow to react, but not in an enduring way.

This, in fact, builds *character*, our second nature, the result of who we are, as we go through life and the various external influences of education and environment, who we have made of ourselves through our willpower and our determinations.

There is a fundamental choice to be made that, as recalled by Augustine in *De civitate Dei*,[5] passes first from the heart of every person and is reflected in the historical processes in which everyone lives. In this possibility for the flowering of himself or herself, the opening up to another, or the selfish closing in the hell of loneliness, it is very easy for them to succumb to disordered and unreasonable love. This do not considered how, in reality, the other becomes part of ourselves, to the point where we can neither hate

[5] «Due amori dunque diedero origine a due città, alla terrena l'amor di sé fino all'indifferenza per Iddio, alla celeste l'amore a Dio fino all'indifferenza per sé. Inoltre quella si gloria in sé, questa nel Signore. Quella infatti esige la gloria degli uomini, per questa la più grande gloria è Dio testimone della coscienza. Quella leva in alto la testa nella sua gloria, questa dice a Dio: Tu sei la mia gloria anche perché levi in alto la mia testa. In quella domina la passione del dominio nei suoi capi e nei popoli che assoggetta, in questa si scambiano servizi nella carità i capi col deliberare e i sudditi con l'obbedire. Quella ama la propria forza nei propri eroi, questa dice al suo Dio: Ti amerò, Signore, mia forza» (Augustine, *De Civitate Dei*, Libro XIV, Cap. 28).

him without even listening to him nor as that part of ourselves that is now rejected maliciously.[6]

Therefore, we must dispel the misunderstanding. Human being, in the profound truth of his identity, has not changed. Also in the age of technology one repeats the eternal struggle between his "own will", as an evil thought to which we are attached and want to stubbornly realize at all costs, and the humble attitude of him who sets himself to listening and to recognize the need to submit to the discipline of a long and liberating journey of purification. This is as long as the entire lifespan, from the destructive tendencies that guarantee us a false personal success, dehumanizing the person and ruining his environment. “Who is bare of virtue, seeks the praise of men” (Dorotheus of Gaza 2014: 105) – wrote Dorotheus of Gaza, a master of discernment, to remind us how easy it would be to slide into inauthenticity, exchanging this praise of others, obtained at any cost, for an illusory self-fulfillment. It is not rather to bury his own conscience, or maybe even daze through technical instrumentalization. It is always Dorothy, in fact, to remind us that, when we step on our conscience. It “like a lamp that gives a dim light because it has dirty oil, begins to show us things in a more confused way, darker and so, gradually. As no one can recognize his face in the murky water caused by a lot of mud, we find ourselves no longer able to perceive any more what it tells us, so much that we almost think that we no longer have it anymore” (Dorotheus of Gaza 2014: 110).

Even in the techno-age, therefore, the fundamental work of man is not any less, namely, that conversion of self, undertaken with more opportunities (Sr. Emaneulle Marie 1998: 105-116) those pathways of awareness of our own connections with other men, called to love and hope in his strength in the face of death, that every person must tread entering into history.

Without hesitation, even in the face of developments that various working groups will make us aware first hand in the field of political, economic, and social organization and of physical and

[6] «La malattia più grave, oggigiorno, non è la lebbra o la tubercolosi ma la solitudine, il sentirsi ignorati, non amati, non voluti. Questa è la causa di tanti disordini, divisioni e guerre che oggi ci affliggono» (Mother Teresa of Calcutta 2016: 28).

biological sciences, we are able to subscribe to the choices, thanks to a more widespread level of education, which is becoming more personal. Almost realizing, without any deterioration in the relativistic subjectivism, all the positivity of the call to responsibility that Søren Kierkegaard set against the disappearance of the subject, operated by the ideological systems of his century: «What matters is to find a truth that would be true *for me*, to find *that for which I want to live and die*» (Kierkegaard 2002: 15).
There is in fact a spiritual level of life, that in times that risk a monistic drift into the neo-materialistic horizon of man reduced to the technical dimension, goes on to rediscover and become cultivated. It, in fact, is burned into us by living life as a progressive decline (in spite of all work to obtain eternal youth!). This invites to avoid leaving oneself captured by the anxiety of his unnatural conservation, helping rather to live it as a progressive journey towards the fullness of love (the New Man of Eph, 4, 2) that, on the one hand, demands fulfillment, but on the other hand, every day of the latter becomes a tangible prolepsis, secure and real.
In the general society, born after the Great World War of 1914-1918 – of which we are retracing the centenary! – when the world stopped being directed by the *elite*, the call to responsibility towards the world as education in the mystery of life, and of humanity, in particular, undoubtedly represented – in our humble opinion – that gathering/reaping and the re-launching of *paidéia*, that makes worthy human history.

References

AGOSTINO, *La città di Dio*, traduzione di D. Gentili, Città Nuova, Rome, 1997.

ALFIERO G. - PICCININ A. (etc), «Democrazia-solidarietà e forme di partecipazione politica» in: SANGALLI S. (a cura di), *Solidarietà e democrazia*, GBP, Rome, 2014.

ALIGHIERI D., *La Divina Commedia. Il Paradiso*, Canto V, vv. 41-42, in: http://divinacommedia.weebly.com/paradiso-canto-v.html.

AMELI F., «Effetto dei nostri atti e responsabilità individuale» in: S. SANGALLI S. (ed.), *Dialoghi di pratica etica*, Luiss University Press, Rome, 2014.

BENANTI P., *The cyborg: corpo e corporeità nell'epoca del post-umano. Prospettive antropologiche e riflessioni etiche per un discernimento morale*, Cittadella Editrice, Assisi, 2012.

BENEDETTO XVI, *Lettera Enciclica Spe salvi*, LEV, Città del Vaticano, 2007.

—— *Caritas in veritate*, LEV, Città del Vaticano, 2009.

BORGNA E., «Fragilità la virtù dimenticata» in: *Avvenire*, martedì 23 agosto 2016.

CASSIANO G., *Le istituzioni cenobitiche*, Qiqajon, Magnano, 2007.

CZERSKI P. , *Millennials manifesto*, in: www.pastebin.org.

S. TOMMASO D'AQUINO, *Le Questioni Disputate. De Malo*, vol VII, (Qu. 16. Ar 3, Responsio), ESD, Bologna, 2003.

DECOSSE D.E., «A primacy of Conscience, Vatican II, and Pope Francis» in: CROWLEY P. s.j. (editor), *From Vatican II to Pope Francis. Charting a Catholic Future*, Orbis Book, NY, 2014.

DOMINGOS P., *L'algoritmo definitivo. La macchina che impara da sola e il futuro del nostro mondo*, Bollati Boringhieri, Torino, 2016.

DONATELLI P., *La vita umana in prima persona*, Edizioni Laterza, Roma-Bari, 2012.

DOROTEO DI GAZA, *Comunione con Dio e con gli uomini. Vita di abba Dositeo, insegnamenti spirituali, lettere e detti*, Introduzione, traduzione e note di Lisa Cremaschi, monaca di Bose, Edizioni Qiqajon, Magnano, 2014.

SR. EMMANUELLE MARIE o.p., «L'accompagnamento spirituale» in: AA.VV., *La attitudine al discernimento*, Ancora, Milano, 1998.

EVAGRIO PONTICO, *Trattato pratico*, Qiqajon, Magnano, 2008.

FACCI M. - VALORZI S. - BERTI M., *Generazione cloud. Essere genitori ai tempi di smartphone e tablet*, Erickson, Trento, 2013.

FRANCESCO, Lettera a Eugenio Scalfari, *La Repubblica*, 11/9/2013.

—— *Laudato Si'*, LEV, Città del Vaticano, 2015.

GALIMBERTI U., *Psiche e techne. L'uomo nell'età della tecnica*, Feltrinelli, Milano, 1999.

GARELLI F., *Piccoli atei crescono. Davvero una generazione senza Dio?* Il Mulino, Bologna, 2016.

GIMMELLI R. - TILLOTTA S., «Solidarietà ed Europa» in: SANGALLI S. (a cura di), *Solidarietà e democrazia*, GBP, Roma, 2014.

GIOVANNI CLIMACO, *La scala del paradiso*, Qiqajon, Magnano, 2005.

HEIDEGGER M., *Holzwege*, Klostermann, Frankfurt a.M., 1963.

IGNAZIO di LOYOLA, *Come si prende una decisione*, Castelvecchi, Roma, 2016.

KIERKEGAARD S., *La malattia mortale. Svolgimento psicologico cristiano di Anti-Climacus*, Edizioni di Comunità, Milano, 1981.

– KIERKEGAARD S., *Aut Aut*, introduzione di R. Cantoni, Mondadori, Milano, 2002.

LAVAZZA A. - MARRAFFA M., *La guerra dei mondi. Scienza e senso comune*, Edizioni Codice, Torino, 2016.

MANCUSO V., *L'anima e il suo destino*, Raffello Cortina Editore, 2007.

NICOTRI F. - DE LUCA E. (etc.), «Centri culturali per un mondo che "soffre di pensiero"» in: SANGALLI S. (a cura di), *Solidarietà e democrazia*, GBP, Roma, 2014.

PESSINA A., *L'io insoddisfatto. Tra Prometeo e Dio*, Vita e Pensiero, Milano, 2016.

PICCININ A. - DE NAPOLI A., «Democrazia e cittadinanza in un mondo globale» in: SANGALLI S. (a cura di), *Sinderesi: fondamenti di etica pubblica*, GBP, Roma, 2012.

PICCININ A., *Religious identity, law and the internet*, LLM by Research, College of Humanity and Social Science, School of Law, University of Edinburgh, 2013.

RUGGENINI M., «L'essenza della tecnica e il nichilismo» in: VOLPI F. (a cura di), *Heidegger*, Editori Laterza, Roma-Bari, 1998.

SANDEL J. M., *Contro la perfezione. L'etica nell'età dell'ingegneria genetica*, Vita e Pensiero, Milano, 2014.

SANGALLI S., «L'etica dei vizi e delle virtù» in: SANGALLI S. (a cura di), *Futuro e Responsabilità. Secondi dialoghi di pratica etica*, LUISS University Press, Roma, 2015.

– (edited by), *A changing humanity. Fast-paced living as a new model of being*, GBP, Rome, 2016.

– «A different age of human history» in: SANGALLI S. (edited by), *A changing humanity. Fast -paced living as a new model of being*, GBP, Rome, 2016.

SEQUERI P. (ed.), *la tecnica e il senso. Oltre l'uomo?*, Glossa, Milano, 2015.

SEVERINO E., *La filosofia futura*, Rizzoli, Milano, 1989.

– *Democrazia, tecnica, capitalismo*, Morcelliana, Brescia, 2009.

– *Capitalismo senza futuro*, Rizzoli, Milano, 2012.

SOLOV'ËV V., *Sulla Divinoumanità e altri scritti*, Jaca Book, Milano, 1971.

– *I tre dialoghi. Il racconto dell'Anticristo*, Marietti, Torino, 1975.

SPADARO A., «Intervista a Papa Francesco», in *La Civiltà Cattolica*, 2013, III, (19 settembre 2013).

ŠPIDLÍK T., *Manuale fondamentale di spiritualità*, Piemme, Casale Monferrato, 1997.

STEIN E., *La struttura della persona umana. Corso di antropologia filosofica.* (Opere complete, 14), Città Nuova – Edizioni OCD, Roma, 2013.

MADRE TERESA DI CALCUTTA, *Amiamo chi non è amato. Testi inediti*, EMI, Bologna, 2016.

TERNYNCK C., *L'uomo di sabbia. Individualismo e perdita di sé*, Vita e Pensiero, Milano, 2011.

TEULINGS C. – BALDWIN R. (edited by), *Secular Stagnation: Facts, Causes and Cures*, CEPR Press, London, 2014.

VACCARO A., *L'ultimo esorcismo. Filosofie dell'immortalità terrena*, EDB, Bologna, 2009.

VANIER J, *La paura di amare*, San Paolo, Milano, 2015.

WILSON E.O., *Il significato dell'esistenza umana*, Le Scienze, 2015.

Lecture n. 2

LAW
Genetics and intellectual property

Dr. Roman Cholij
Research Associate
Von Hügel Institute
University of Cambridge (UK)

Introduction

On 23 September 1999 Pope, now Saint, John Paul II wrote "the Catholic Church ... has a moral vision of what the good of individuals and of the human family demands. She has consistently taught that there is a "social mortgage" on all private property, a concept which today must also be applied to "intellectual property" and to "knowledge". The law of profit alone cannot be applied to that which is essential for the fight against hunger, disease and poverty" (Message of the Holy Father to the Group Jubilee 2000 Debt Campaign).

Ten years later, Pope Benedict XVI, in his encyclical *Caritas in Veritate* (29 June 2009, §22), in the context of the ever growing distance between rich and poor, developed and developing countries, noted "On the part of rich countries there is excessive zeal for protecting knowledge through an unduly rigid assertion of the right to intellectual property, especially in the field of health care".

What these two quotes demonstrate is that Intellectual Property is now on the 'radar screen' of the Catholic Church, now part of the language of Magisterial teaching and of Catholic Social Doctrine.

In this presentation I would like first to examine the role of intellectual property from a Christian ethical perspective involving the conservation and the equitable exploitation of Genetic Resources in the context of biodiversity, Traditional Knowledge ('TK') and least developed countries of the world. Our concern will be primarily focused on social justice issues, but from the perspective of the benefits and problems of the current IP system.

Following on this, and with a slight shift in focus, we will look at some particular areas of intellectual property protection through patents that are part of a larger global economic, legal and cultural revolution constituted by biotechnology, in particular molecular or genetic biotechnology with its multiple applications. Such applications include those in agriculture, healthcare, life style products, human enhancement and so-called assisted reproduction technologies (ARTs). Many of the new technologies and their applications raise serious bioethical issues, especially when applied to human beings. Some of these issues of morality will be touched upon in this paper, but principally only in so far as they are relevant to intellectual property law. A more thorough treatment of bioethics would lie outside the remit of this paper.[1]

Structure of this paper

After a brief explanation of the nature of intellectual property for non-specialist listeners/readers, the following topics related to genetics will be considered.

The first half of this paper, entitled **'The Protection of Genetic Resources through Patents and its Global context'** will be devoted to social issues that surround the use and abuse of natural genetic resources and indigenous traditional knowledge, and will look at the role of patents. Genetics in this instance will be understood as living matter encoded by genes from whole organisms to microbial and sub-cellular components. This is an area in which the Holy See itself has been engaged as it involves issues of social justice in the developing world.

Part II of this paper, entitled **'Life patents',** will look at the biotechnology industry, giving some history and background, the role of patents in building up this industry, looking at what patent ownership really means and whether we can really speak, as some do when

[1] Pope John Paul II, conscious of the importance of the new field of bioethics, established in 1994 the Pontifical Academy for Life as a forum for Catholic engagement in these matters The Pontifical Academy exists for the promotion and defence of human life, especially regarding bioethics as it regards Christian morality. Its statute was defined by the motu proprio Vitae Mysterium of Pope John Paul II on 11 February 1994. See Pontifical Academy for Life [website]. As part of its activities the Workshop "Technology and Human Generation. 30 Years after *Donum Vitae*", was held on March 2, 2017 at the New Synod Hall in the Vatican City.

attacking life patents, of 'ownership of life' through patents. We will look at objections to patenting, patent law and morality, and will draw out the conclusion that despite the various controversies that will be mentioned, intellectual property protection through patents and genetic research and experimentation that is beneficial to mankind can be fully embraced by a conscientious Christian, and falls within the parameters of the Social Doctrine of the Church.

Part I

1. The Protection of Genetic Resources through patents and its Global context

1.1 What is intellectual Property?

Intellectual property (IP) refers to creations of the mind, such as inventions; literary and artistic works; designs; and symbols, names and images used in commerce.
IP is protected in law by, for example, patents, copyright and trademarks, which enable people to earn recognition or financial benefit from what they invent or create. By striking the right balance between the interests of innovators and the wider public interest, the IP system aims to foster an environment in which creativity and innovation can flourish.
Copyright is a legal term used to describe the rights that creators have over their literary and artistic works. Works covered by copyright range from books, music, paintings, sculpture and films, to computer programs, databases, advertisements, maps and technical drawings.
A patent is an exclusive right granted for an invention. Generally speaking, a patent provides the patent owner with the right to decide how - or whether - the invention can be used by others. In exchange for this right, the patent owner makes technical information about the invention publicly available in the published patent document.
A trademark is a sign capable of distinguishing the goods or services of one enterprise from those of other enterprises. Trademarks

date back to ancient times when craftsmen used to put their signature or "mark" on their products.[2]
The area of intellectual property most important to our purposes is patents.

1.2 The patenting of Genetic Resources

Our first consideration is the role of patents in the legal protection of Genetic Resources ('GR'), a topic very closely related to global biodiversity and to natural traditional medicine.
Firstly, a note on terminology. Genetic Resources can be defined as 'Genetic material of actual or potential value, this genetic material being any material of plant, animal, microbial or other origin containing functional units of heredity". [3]

Biodiversity and the WHO

Because of its direct link to health issues, the World Health Organisation ('WHO') has taken a particular interest in biodiversity. On the Organisation's website, under the Heading of Climate change and human health' the following is said about biodiversity: "...biophysical diversity of microorganisms, flora and fauna provides extensive knowledge which carry important benefits for biological, health, and pharmacological sciences. Significant medical and pharmacological discoveries are made through greater understanding of the earth's biodiversity. Loss in biodiversity may limit discovery of potential treatments for many diseases and health problems.
There is growing concern about the health consequences of biodiversity loss and change. Biodiversity changes affect ecosystem functioning and significant disruptions of ecosystems can affect life sustaining ecosystem goods and services. Biodiversity loss also means that we are losing, before discovery, many of nature's chemicals and genes, of the kind that have already provided humankind with enormous health benefits."

---

[2] This summary has been taken from the *World Intellectual Property Organization* (2001)
[3] Definition used by the Convention on Biological Diversity, Article 2

Traditional medicine continues to play an essential role in health care, especially in primary health care. Traditional medicines are estimated to be used by 60% of the world's population and in some countries are extensively incorporated into the public health system. Medicinal plant use is the most common medication tool in traditional medicine and complementary medicine worldwide. Medicinal plants are supplied through collection from wild populations and cultivation. Many communities rely on natural products collected from ecosystems for medicinal and cultural purposes, in addition to food.
Although synthetic medicines are available for many purposes, the global need and demand for natural products persists for use as medicinal products and biomedical research that relies on plants, animals and microbes to understand human physiology and to understand and treat human diseases".[4]

Contribution of Traditional Knowledge to new drugs

It has been estimated that between 33% - 50% of modern pharmaceutical preparations were derived originally from plants.[5] A recent testimony of the on-going importance of plant-based medicines is the award in 2015 of the Nobel Prize for Medicine to the first Chinese woman ever to receive such an award Tu Youyou for the discovery of the most successful anti-malarial drug on the market. Tu's team found information in a 1,600 year old text that people in 400 AD used sweet wormwood (Artemisia annual.), which is known in Chinese as qinghaosu, to treat malaria with some success. They used this information to create the antimalarial drug (Lee 2015).
Many (but not all) of the genetic resources that possess significant economic and social value are located in territories that have been

[4] World Health Organization website: www.who.int/en/

[5] Some prominent examples include digitalis, a popular cardiac medication, identified as the active component of the foxglove leaf; morphine and codeine, which alleviate pain, derived from cultivated opium poppy; and atropine, for disorders involving the autonomic nervous system, from the nightshade plant. The anticancer drug Taxol was derived from the bark of the Pacific yew tree, and Aspirin was isolated from willow bark (Abbott 2014).

inhabited since time immemorial by native communities within the jurisdiction of countries different from those where the further processing and commercial exploitation of the genetic material takes place, namely in the developed world and often in the United States (by pharmaceutical companies in the case of drugs based on Genetic resources, such as Pfizer, Roche, Novartis, Amgen, Merck & Co. and others).[6] The exploitation of these drugs however is always preceded by patent protection. This protection is usually world-wide, especially in the case of drugs with significant therapeutic potential and would usually also include the country of origin of the GR, such as Brazil or Peru (in the case of Amazonian forest derived products), thereby preventing the manufacture of generic versions of the patented drug.

Genetic resources linked to traditional knowledge held by native communities – the indispensable 'clue' to the potential proven medicinal uses of a herb, for example – can potentially save private companies large amounts of money in research and development costs. The genetic resource could be a plant, fungus, microorganism or animal poisons and toxins (such as from snakes, frogs and toads) from which substances could be extracted and isolated. The search for new drugs based on GR and TK is known as 'bioprospecting' by its protagonists. However, the unilateral pocketing of the profits from these resources and knowledge, without benefit sharing with those providing the original GR and TK, is also considered by many non-commercial groups as a misappropriation, or even as 'biopiracy'. This situation is essentially only made possible by the patenting system which excludes unauthorised use by non-patent holders, giving the holder an ownership and price monopoly.[7] Those who first

[6] These are some of the multi-billion dollar companies who can make billions of dollars (not just millions) on block-buster drugs.

[7] Biopiracy is actually a complex issue. For the genesis of the term see ETC Group [website]: "Biopiracy, a term originally coined by ETC Group, refers to the appropriation of the knowledge and genetic resources of farming and indigenous communities by individuals or institutions that seek exclusive monopoly control (patents or intellectual property) over these resources and knowledge. ETC Group believes that intellectual property is predatory on the rights and knowledge of farming communities and indigenous peoples". The literature on biopiracy is very extensive. Some examples include: Robinson 2010; Nagan et al. 2010; Chen 2006; Ho 2006.

made possible the success of large corporations, by 'pointing the way', are generally the poor and marginalised native or aboriginal peoples of various developing countries (in particular from South America, Africa and India).[8]

In 2008 Indigenous peoples from around the world gathered together and made a statement for the G8 nations who had gathered for a summit in Hokkaido, Japan, asking them "to stop the theft and piracy of our traditional indigenous knowledge, traditional cultural expressions [and] bio-genetic resources including our human genetic resources, by biotechnology corporations, cultural industry and even by States and individual scientists and researchers."[9]

Apart from the aspect of non-benefit sharing, a particular issue raised by concerned parties and affecting a number of the patents contributing to the divide between rich and poor, developed and least developed/developing countries, is that a number of these patents are 'bad' patents or patents that should never have been granted based on technical criteria, especially that of 'lack of novelty'. Where TK is involved, this is already knowledge in the 'public domain' even if held by a few, and as such undermines one of the essential aspects of a patent, namely 'novelty' (we will discuss this further below).

1.3 Two case studies – Neem and Turmeric

An oft-cited example of 'biopiracy', or misappropriation of Traditional Knowledge in association with a genetic resource, is that involving Neem (Azadirachta indica). Neem ('Indian Lilac') is a plant that is endemic to the Indian sub-continent and traditional medicine practitioners for its various therapeutic properties. The Neem tree has been the subject of a considerable number of patents, with more than 40 in the US alone and at least 150 worldwide. Traditional Knowledge was used as a starting point in each case. This appropriation (or misappropriation) of TK together with a GR has aroused considerable controversy, especially in

[8] We should also point out, for the sake of balance, that many 'leads' do not go anywhere or may be only very incidental to the development of a future drug or some other useful product based on the original GR and/or TK.

[9] Nibutani Declaration of the 2008 Indigenous Peoples Summit in Ainu Mosir §17

India.[10] One patent, entitled 'a method for controlling fungi on plants by the aid of a hydrophobic extracted neem oil', owned by a US company, was revoked at the European Patent Office when challenged based on evidence supplied to show that there was a lack of novelty and inventive step (EP 436257B1).[11]
Another patent, (US5401 504) 'Method of promoting healing of a wound by administering turmeric to a patient afflicted with the wound', owned by the University of Mississippi , involved Turmeric, a herbaceous plant native to tropical South Asia, likewise known through Traditional Knowledge to have healing properties. This was similarly challenged successfully, based on evidence of prior art that consisted in TK. In this case an ancient Sanskrit text was used by the Indian government to show lack of novelty (the US Patent Office not accepting non-written oral knowledge).[12]
What the above examples show, of which there are plenty more documented examples, is that patents can be, and have been, awarded erroneously based on patent criteria that have not been properly applied by examining authorities.[13]

---

[10] The search term 'Neem' entered into the database of Espacenet of the European Patent Office produced 560 patents worldwide, the first entries including herbal topical formulation, stretch mark preventer, method for controlling and/or repelling pests etc.

[11] The abstract reads: *A novel insecticide and foliar fungicide derived from a neem seed extract comprising neem oil which is substantially free of azadirachtin and salannin, said neem oil being prepared by extracting dried, coarsely ground neem seeds with a non-polar, hydrophobic solvent to obtain a neem oil extract, and then removing the solvent to obtain the neem oil. These neem oil pesticides exhibit the ability to repel insects from plant surfaces, prevent fungal growth, and kill insect and fungal pests at various life stages.*

[12] There are a number of dubious applications for patent protection, including from Korea to protect such 'inventions' as 'Chocolate containing Turmeric and method for preparing the same', 'Turmeric Vinegar' and similar.

[13] Erroneous patents include those on the enola bean, the rosy periwinkle, hoodia, quinoa, ayahuasca, maca etc. See for example Tzen Wong – Dutfield 2011. There have also been in recent years efforts by the World Intellectual Property Organisation and other organisations to provide appropriate database systems and mechanisms whereby TK is recorded which all patent offices can access to prevent the granting of such erroneous patents.

1.4 Using the patent system to promote social justice?

Because of the powerful forces in play and competing interests, the relationship of genetic resources and TK to intellectual property protection has been one of the most complex, controversial and dynamic issues on the agenda of multilateral diplomatic deliberations during the past decade.

For example, the Convention on Biological Diversity ('CBD'), an international legal instrument for the conservation and sustainable use of biological diversity, presented for signature at the Rio 'Earth Summit' in 1992, required that access to genetic resources be on the basis of prior informed consent, and on mutually agreed terms that provide a fair and equitable sharing of the benefits arising out of the utilization of the genetic resources.[14] Related to this instrument and developed by the Conference of Parties ('COP') to the CBD are the Bonn Guidelines that invite parties to the Convention to encourage disclosure of the country of origin of GR and TK in the patent applications of those who are set to commercially benefit from the patent. Concerns had been previously expressed on the part of NGOs and others that contractual arrangements between, say indigenous peoples and foreign companies, were often quite ineffectual in securing the objectives of the Convention, for example because of the disparity in bargaining power of the parties. Mandatory disclosure in a patent application, with information that could then be checked, would also show whether or not effective arrangements to benefit the holders of GR and TK, usually the poor and marginalised, were in place.

The World Intellectual Property Organisation (WIPO), a UN agency based in Geneva, has been working with these issues, and trying to resolve conflicting interests, mainly between developing and developed countries, in its *Inter Government Committee on Intellectual Property, Genetic Resources, Traditional Knowledge and Folklore* (IGC) since 2001. The objective is to arrive at an international legal instrument on the patenting of genetic resources that is consistent with the CBD objectives. This work is still ongoing.[15]

[14] Article 1

[15] The latest version (on writing) of the consolidated document of the text being negoti-

1.5 Role of the Holy See

From the inception of the IGC , the Holy See, through its permanent observer mission to the UN in Geneva, has taken great interest in and contributed actively to the discussions at WIPO regarding the adaptation of the international patent system to help secure the ends of the CBD, including supporting mandatory disclosure of the source of GK and TK. In a masterly short document submitted on April 26, 2001, the Holy See sets out key principles of Catholic Social teaching that apply to the complex area of intellectual property in relation to securing effective rights for the holders of GR and TK, emphasising the proper balancing of interests in view of higher principles (World Intellectual Property Organization 2001). A body of interventions to this and other international organisations, in particular the World Trade Organisation, likewise emphasise key aspects of social justice in the implementation of intellectual property, in particular patent, laws.[16] A point constantly emphasised is the social function of intellectual property law, or its ordering towards the common good, which should be the end of all human law. In this vision the right to intellectual property which can be considered a special type of private property is not an absolute right but must be tailored towards the needs of wider society. And the preferential option for the poor is a particularly important guiding principle directing the resolution of conflicts regarding the poor. We will return to these points shortly.

ated between governments was published June 14, 2016: http://www.wipo.int. The long delay in issuing an agreed document appears to be due to entrenched and uncompromising opposite views on the nature and function of the patent system. Top users of the patent system, such as the United States (who incidentally is the sole nation not to have ratified the CBD), under pressure from large corporations have resisted any change in the patent system that could involve more cost and be more burdensome to use.

[16] Some of these documents are conveniently gathered by the *Caritas in Veritate* Foundation Caritas (2013).

1.6 Conflict between poor and rich countries

The tensions that exist among government delegates at WIPO also reflect broader tensions that exist in relation to international patent laws as such. Developed countries see patent laws, as with the rest of intellectual property laws, primarily in terms of the economic benefits that they convey. Inventors should be rewarded for their labour by (temporary) monopolies on patented goods or processes; companies who acquire patents should be able to charge premium prices to recoup the costs of research and development and the marketing of goods as well as making profits that encourage future investments. There is also a benefit to society, they claim, in terms of making public new knowledge which, even though temporarily protected in terms of economic exploitation (generally 20 years), contributes to the general scientific knowledge pool and encourages further research by others.
Developing countries, on the other hand, whilst not rejecting the economic rationale of IP, also see intellectual property laws in terms of their contribution to societies as a whole not just to private companies and large international corporations driven by the push for profit. If IP laws affect society in a negative way, for example by impeding access through high prices to important medicines, such as against HIV/Aids in many African countries, whilst blocking the possibility of importing or manufacturing generic drugs, then other considerations of the benefits of IP are totally one-sided or theoretical at best.
Traditionally IP laws have been very flexible in terms of what could be given patent protection and what could not. Large pharmaceutical companies, for example, could not hold patents in countries like India or African countries for healthcare products such as drugs, nor for other things, e.g. seeds essential for life or for agricultural economy. All governments had the powers to regulate what was beneficial and what was not for their countries and citizens.
Even European countries, earlier in their history and industrial development, avoided granting patents, especially to foreign companies, so as to allow their industries to develop through 'copy-

ing'. Italy, Japan and Switzerland, for example, did not allow pharmaceutical patent protection until development reached a point where the per capita income was $20,000. And yet today these same countries expect African countries, and others, with a per capita income perhaps of just $594, to adopt the same intellectual property laws, allowing foreign companies to have full monopoly rights in their country.[17]

1.7 Intellectual Property as an instrument of Trade – the TRIPS agreement

What has changed, in terms of the social history of intellectual property, is that primarily through US corporation pressure (but opposed by developing countries) all signatories to the GATT free-trade agreement administered by the WTO, allowing in principle poor countries to trade with richer countries without protectionist high tariffs blocking their exports, have been required to sign a separate intellectual property agreement known as TRIPS (Agreement on Trade-Related Aspects of Intellectual Property Rights).[18] The TRIPS agreement introduced intellectual property law into the international trading system for the first time, removing discretionary powers of governments to exclude patenting in key areas such as healthcare

TRIPS requires WTO members to provide copyright rights, covering content producers including performers, producers of sound recordings and broadcasting organizations; geographical indications, including appellations of origin; industrial designs; integrated circuit layout-designs; patents; new plant varieties; trademarks; trade dress; and undisclosed or confidential information. We might note in passing that the major economic

[17] This figure is taken from the International Monetary Fund for the Central African Republic. Many other countries in Africa have under $1000 OR $2000 as GDP per capita. See Wikipedia [website]. See also United Nations 2010.

[18] This was done at the end of the Uruguay Round of the General Agreement on Tariffs and Trade (GATT) in 1994. It was actually the senior management at Pfizer in the early 1980s, who mobilized corporations in the United States and made maximizing intellectual property privileges the number one priority of trade policy in the United States. See Braithwaite – Drahos 2000.

interests that dictated the forcing through of this agreement, and who had the most direct stake in TRIPS, included some of the world's largest and most profitable industries: the US $650 billion per year global pharmaceutical industry... the commercial seed industry worth an estimated $21 billion per year), and the global software and entertainment industries (worth an estimated $800 billion per year). In many of these industries, a handful of companies monopolized markets and their business models depended on securing IP protection. The playing field was uneven. Throughout the TRIPS negotiations, developed country corporations, research centres and individuals together held over eighty per cent of the world's IP rights. Developing or least developed countries were in principle now required to allow IP protection to foreign companies regardless of whether this would benefit their country. This was the price to pay for free trade (Deere 2009). [19]
This talk is not about TRIPS, but it is important to have it in mind as context to understand some of the broader issues of IP policy and politics, as well as to understand why the Holy See has also become involved in matters which, perhaps prior to the 1990's were generally left to those with particular technical knowledge, in particular of intellectual property law.[20]

1.8 The social obligations associated with Intellectual Property rights – Church Teaching

The developed nations who are the prime users of the international IP system, it could be argued, have a mind-set in which social benefits or common good aspects of patents are secondary to that of securing legal protection for financial investment, and this mind-set also finds expression in the resistance to modify patent law structures for the sake of the poor with regard to genetic resource exploitation. For this reason it is salutary to proclaim to the corporate world the basic messages of Catholic social doctrine.

[19] Dr Carolyn Deere is Senior Researcher at the Global Economic Governance Programme at the University of Oxford , founder and Chair of the Board of Directors of Intellectual Property Watch – the world's leading news service on international intellectual property policy debates. See www.ip-watch.org
[20] See also my discussion in *Caritas in Veritate* Foundation (2013).

Universal destination of goods

The universal destination of goods, for example, emphasises that the right to ownership of property – tangible or intangible like intellectual property – is not absolute and untouchable. It must be oriented to the common good:

According to the *Compendium of the Social Doctrine of the Church*, intellectual property, like all property, needs to be oriented "to an economy of service to mankind, so that...[it] contribute[s] to putting into effect the principle of the universal destination of goods." ... "The issue of ownership and use of new technologies and knowledge – which in our day constitute a particular form of property that is no less important than ownership of land or capital – becomes significant in this perspective. These resources, like all goods, have a universal destination; they too must be placed in a context of legal norms and social rules that guarantee that they will be used according to the criteria of justice, equity and respect of human rights." (*Centesimus Annus*, n. 32).

This fundamentally important doctrine of the universal destination of goods provides a particularly Christian qualifier on the exercise of IP rights that emphasises the 'social dimension' of intellectual property and possession of knowledge goods. It likewise provides an important orientation for matters touching TRIPS and other international negotiations such as on patent protection of Genetic Resources at WIPO. The Compendium continues: "New technological and scientific knowledge must be placed at the service of mankind's primary needs, gradually increasing humanity's common patrimony. Putting the principle of the universal destination of goods into full effect therefore requires action at the international level and planned programmes on the part of all countries. It is necessary to break down the barriers and monopolies which leave so many countries on the margins of development, and to provide all individuals and nations with the basic conditions which will enable them to share in development." (*Compendium*, n. 179).

This Christian social perspective or dimension on intellectual property has also been referred to as its 'social mortgage' or encumbrance (*Sollicitudo Rei Socialis*, n. 42).[21] "The intellectual property rights system exists not just to protect creative and innovative impetus but also and primarily to serve the common good of the human family. As a universal common good, intellectual property demands that control mechanisms should accompany the logic of the market."[22]

Pope Francis, in his encyclical Laudato Si', *On Care for our Common Home* has also re-stated this principle in a powerful and direct way.[23]

In its first statement to the IGC for Genetic Resources, Traditional Knowledge and Folklore, the Holy See expressed the principle thus:

> "*The raison d'être of intellectual property protection systems is the promotion of literary, scientific or artistic production and inventive activity for the sake of the common good. That protection officially attests the right of the author or inventor to recognition of the ownership of his work and to a degree of economic reward, at the same time as it serves the cultural and material progress of society as a whole. The ultimate cause that intellectual property protection works for is the recognition of the dignity of man and his work, in its double dimension, namely as a medium of expression and growth of the individual personality and as a contribution to the common good.*"[24]

---

[21] This language is borrowed from residential property sales in which the buyer borrows some of the funds needed for the purchase. For an analysis of this concept in the thought of John Paul II, see O'Boyle 2012; Carey 2007, and the Intervention by his Excellency Mons. Diarmuid to the Plenary Council of the World Trade Organization on Trade-Related aspects of Intellectual Property Rights. Wednesday, 20 June 2001, par 11: "The Holy See, consistent with the traditions of Catholic social thought, underlines that there is a 'social mortgage' on all private property, namely, that the reason for the very existence the institution of private property is to ensure that the basic needs of every man and woman are met and sustained. This 'social mortgage' on private property must also be applied today to 'intellectual property' and to 'knowledge' (John Paul II, Message to the "Jubilee 2000 Debt Campaign" Group, September 23, 1999). The law of profit alone cannot be applied to that which is essential for the fight against hunger, disease and poverty. Hence, whenever there is a conflict between property rights, on the one hand, and fundamental human rights and concerns of the common good, on the other, property rights should be moderated by an appropriate authority, in order to achieve a just balance of rights."

[22] Intervention by the Holy See at the 5th Ministerial Conference of World Trade Organisation (WTO), Address of Msgr. Frank Dewane, Cancun, Mexico, Saturday, 13 September 2003

[23] For example at n. 93 'The Common Destination of Goods'

[24] *WIPO/GRTKF/IC/1/7*, 2.

Also in line with the principles of solidarity and the preferential option for the poor ,the Holy See's position as a Permanent Observer to the UN is to align itself with the concerns of the least developed and developing nations and support their proposals to achieve the best possible mechanism in justice from the patent system for their role in providing GR and TK to other parties, institutions and corporations (which may include mandatory disclosure of the origin of GR and TK to help ensure free consent and appropriate benefit sharing). At the First session of the *Intergovernmental Committee on Intellectual Property and Genetic Resources, Traditional Knowledge and Folklore*, held in Geneva, April 30 - May 3, 2001, the Holy See stated:

> "*With regard to the specific work of the Intergovernmental Committee, it would be desirable that new legal machinery be successfully created that is fully integrated in and consistent with the international provisions currently in force, and imposes on the legislation of all Member States of WIPO certain minimum protection requirements for those sectors whose rights and interests are not fully catered for in systems now in force. In the case of biological resources, the Holy See considers that the proposed tasks … as a whole should result in the drafting of guidelines that guarantee the following objectives:*
>
> a) acceptance of the institution of the informed, free consensus of persons, peoples and States as a prerequisite of patenting
> b) achievement of equitable economic participation of native populations in the benefits deriving from the commercial exploitation of biological resources… [25]
> c) Fifteen years after this was written this objective has still not been guaranteed.

[25] WIPO/GRTKF/IC/1/7, 10. Point d) in the same paragraph also expresses a common social concern: "assurance that patents for biological discoveries do not constitute an undue obstacle to subsequent research and scientific teaching." Point c) mentions human genetic resources.

PART II

2. Life Patents

2.1 Why patent protection is sought

The rest of this paper will deal with issues that are not primarily related to developing countries and issues of social justice. Our next topic is a consideration of what it means to have a patent on a living organism such as a plant or animal or on something that has been genetically altered. Does a patent confer ownership to a private individual or company on something that perhaps should be considered part of the common heritage of mankind? The legislation of TRIPS allows nations – with developing nations principally in mind – to exempt themselves from the requirement of providing patent protection over plants and animals, which is allowed in the US, Europe, Japan and several other countries.[26]
The explosion of the biotechnology industry since the 1970's and 1980's has brought with it the need to secure protection against unauthorised copying or reverse-engineering against the massive investments of larger corporations in research & development and seeing products through the regulatory stages before going to market. It is estimated, for example, that to bring a new plant biotechnology derived trait to market the cost involved is around $136 million and takes on average 13 years (McDougall 2011).
A recent report in the Scientific American states that the estimated cost for introducing a new pharmaceutical drug to market is around $2.5 billion (Mullin 2014). A new drug might be derived from a natural genetic resource that has been subsequently modified.

[26] TRIPS Article 27.3(b) states in part Members may also exclude from patentability: (a) diagnostic, therapeutic and surgical methods for the treatment of humans or animals; (b) plants and animals other than micro-organisms, and essentially biological processes for the production of plants or animals other than non-biological and microbiological processes. However, Members shall provide for the protection of plant varieties either by patents or by an effective sui generis system or by any combination thereof… It may be noted that the text does not specify if genes can be exempted.

2.2 Some history of the biotechnology industry

The biotechnology industry as such is not in fact a modern phenomenon. Beer brewing and bread making, for example, use natural microorganisms in fermentation and the techniques involved could be considered to be a form of biotechnology. We know that the earliest agricultural societies, such those from Mesopotamia – Sumerians, Babylonians and Akkadians – brewed beer. Carvings on the walls of Egyptian tombs also show that beer making was an acquired technology. Christian monks for many centuries have also brewed beer with sophisticated techniques – sometimes being the only alternative to polluted water sources. Perhaps this helped keep monks in good humour!

Careful selective breeding of plants, including through grafting and creating hybrids, can also be considered a proto-form of biotechnology. Work begun in 1908 in the USA eventually led to a form of hybrid corn that revolutionised US agriculture, rapidly increasing corn yields. This was of course a very positive development, but a side effect was that farmers could not replant corn seed as the offspring of hybrid corn varieties do not have the same properties as their parents. This meant a dependency on commercial seed producers. There was therefore a natural in-built form of protection that did not require IP rights to protect the breeder's investment.

New hybrid forms of other crops, such as sugar-beet, maize, vegetables such as potatoes, fruits and flowers were also biotechnologically engineered in the early years of the twentieth century and increasingly following the Second World War. In the 1960's, especially with the introduction of high yielding semi-dwarf varieties of rice and wheat (with the backing of the Ford and Rockefeller foundations in the US) this form of plant genetic modification through breeding techniques led to what has been called the 'Green Revolution' (Stevens 2016). Effectively the genetic makeup of these live organisms were being changed by man, although in what we can call a 'natural' manner that did not involve interventions at a molecular level, in particular genetic recombinant technology which we will further consider below.[27]

[27] We should note also that plants are subject to 'natural' genetic mutations.

We can also say the same thing with modifications of animals – domestic animals for example have all been modified genetically through breeding.
From the time of Louis Pasteur onwards, and with the discovery of penicillin, antibiotics have also been produced from bio-techniques involving fermentation.
The explosion of the biotech industry really occurred with the advent of genetic recombinant technology which involved cutting or splicing a piece of DNA from a plasmid of a bacterium (a ring of double-stranded DNA separate from the main bacterial chromosomes) and 'sticking' a similarly spliced piece of DNA – taken from anywhere – and then multiplying this in a culture. The idea of doing this – combining two necessary research techniques – came from a meeting of a young molecular biologist from the University of California, Herbert Boyer, with a geneticist, Stanley Cohen, from nearby Stanford University in November 1972. A walk along Waikiki beach in Honolulu, followed by a beer and sandwiches in a deli restaurant resulted in a collaboration that would lead to the new era of the biotechnology – that of genetic engineering. The heart of the new biotechnology industry that would be created rapidly as a result of this breakthrough was Silicon Valley in California, the home also of high-tech and computer companies (Stevens 2016). Hallam Stevens, who has documented this early history writes:

> "*Cohen and Boyer's discovery formed the basis for what became the biotech industry. Globally the combined revenue of this industry was over $250 billion in 2013…But the industry continues to grow rapidly: it includes not only the sale of drugs and therapeutics, but also genetic tests, personalized medicine, the sale of technologies such as DNA chips and sequencing machines, trade in patents, the running of global clinical trials, and the use of biotechnologies in agriculture*" (Stevens 2016: 47).

2.3 Biotechnology and Patents

Patent law: some basics.

2.3.1 Essential requirements

Article 27(1) of TRIPS establishes that patents should be available for all inventions, whether products or processes, in all fields of technology, provided that they are new, involve an inventive step (or are 'non-obvious to someone skilled in the art') and are capable of industrial application (or, as in the US are 'useful' – i.e. not just knowledge for its own sake).[28]

It is accepted that the basic aim of a patent is to provide incentives to invent or innovate, thus contributing to the economy and to science. If you invented a new type of mouse-trap or toothbrush (that cleans upper and lower teeth at the same time), and you wished to sell your invention to make a living, once it is on the market, and especially if it is successful, others could just reverse-engineer or copy the product. A patent allows the inventor to profit from his invention without fear that someone can copy and 'get away with it'. However, in order to deserve this recognition there are technical requirements. The invention must be truly new. If there is a disclosure of something very similar, even if it is in another language and in a journal or leaflet at the other end of the world, then the very possibility of having seen and copied it (more real today than in the past with the Internet) is enough to prevent a patent monopoly. Also the invention can't be trivial. If you took a mouse trap of a certain size, and just changed its dimensions so that you could also catch rats and not just small mice, or manufactured your toothbrush using a different material, then this would not have required much ingenuity or skill. Such developments would be regarded as obvious and not deserving of

[28] TRIPS Art 27 (1.) reads: "Subject to the provisions of paragraphs 2 and 3, patents shall be available for any inventions, whether products or processes, in all fields of technology, provided that they are new, involve an inventive step and are capable of industrial application. Subject to paragraph 4 of Article 65, paragraph 8 of Article 70 and paragraph 3 of this Article, patents shall be available and patent rights enjoyable without discrimination as to the place of invention, the field of technology and whether products are imported or locally produced."

a patent monopoly. Lastly, your invention must be useful or have some application in industry. This would exclude, for example, a new design to make something look more attractive, like a kettle, or a cartoon character. These could be protected under separate categories of IP, namely designs and copyright. Although TRIPS does not spell out what is necessarily excluded, other jurisdictions do. For example in Europe the European Patent Convention specified, in Article 52(2), that the following cannot be regarded as inventions:

(a) discoveries, scientific theories and mathematical methods;
(b) aesthetic creations;
(c) schemes, rules and methods for performing mental acts, playing games or doing business, and programs for computers; [these last two make Europe different from the US where these are allowed]
(d) presentations of information.

2.3.2 The meaning of patent ownership

It is important, especially in the context of our discussion on 'life' patents, to understand what the patent holder actually has in his possession. What does he 'own'?

TRIPS spells out in very exact terms that what the patent owner holds is essentially a negative right in law to exclude others from, in the case of products, making or using the invention, and certain other acts like importing without authorisation (which if allowed might be allowed under a licence). This is aimed primarily at commercial competitors.[29] It is a right that can also be sold or assigned. In other words, it is not a piece of personal property like owning a car, computer, cat or house plant, but a right in intangible property. But nor does this mean that it is ownership of what

[29] Article 28.1 reads: A patent shall confer on its owner the following exclusive rights: (a) where the subject matter of a patent is a product, to prevent third parties not having the owner's consent from the acts of: making, using, offering for sale, selling, or importing for these purposes that product; (b) where the subject matter of a patent is a process, to prevent third parties not having the owner's consent from the act of using the process, and from the acts of: using, offering for sale, selling, or importing for these purposes at least the product obtained directly by that process.

is claimed in the patent. In terms of, for example, patented seed, the ownership rights of a box of seed would be transferred from the supplier to the buyer who could also then sell this on, but the package of IP rights in the seed would remain with the patent owner – rights to exclude - and would prevent, for example, next generation re-seeding requiring the user (farmer) to buy new seeds the following year. Even the patent owner may be excluded from 'use' of his own patent in an embodied product if, for example, it was against the law to do so – like exploding a new bomb or, more commonly, if it were an improvement on another patented product in which case that patent owner's authorisation would first be needed.

As we will see shortly, theological and ethical discourse on life patent 'ownership' needs, therefore to be tempered, bearing in mind this distinction, especially where expressions such as private companies 'owning life' or having 'ownership over life' are used, and the corollary of the (rather ambiguous) concept of 'commodification of life,' referring to the commercial transaction side of patent rights on living organisms.[30]

2.3.3 A limited monopoly

Another important point about patents is that they are time-limited, ordinarily to twenty years from first filing of the patent application.[31] This is in order to create a balance that respects the theoretical 'social bargain' that is implicit in the granting by the Government of a patent. The inventor or investor has the monopoly reward for inventing and investing in new technology and disclosing in the patent application useful information about the invention which allows other parties to avoid repetition of research and to further develop the inventive concept.[32]

---

[30] This concept may be more justly applied to matters such as human and human body part trafficking.

[31] Trips Article 28 states: "The term of protection available shall not end before the expiration of a period of twenty years counted from the filing date."

[32] They might not be able to actually use the new invention without a licence if it embodies the technology of the earlier patent during its lifetime.

2.4 Can living organisms be considered 'inventions'?

Inventions traditionally are thought of in terms of mechanical or engineered objects, or perhaps new chemical or pharmacological combinations that lead to new therapies. The extension of the concept to living organisms has challenged the IP system, but has also shown that the system has developed primarily as a servant of industry and the economy, and less as a tribute to creativity.
The rationale for the extension of the IP Regime into the realm of living organisms and the field of bio- and genetic technologies, is expressed in the words of the EU Biotechnology Directive, in its introduction or 'whereas' preamble:

(1) Whereas biotechnology and genetic engineering are playing an increasingly important role in a broad range of industries and the protection of biotechnological inventions will certainly be of fundamental importance for the Community's industrial development;
(2) Whereas, in particular in the field of genetic engineering, research and development require a considerable amount of high-risk investment and therefore only adequate legal protection can make them profitable;

As with the US, in the EU biotechnology is recognised as big business which 'contributes to industrial development' but which involves 'high-risk investment'. Patent law has business as its primary focus.
The expansion of IP on a global level into the arena of living organisms actually happened in an almost surreptitious way – originating in the US through judicial interpretation of the law, crossing into Europe and now affecting the whole world through Article 27, 3 (b) of TRIPS. The patent regime has now extended itself to matters that affect food security and to ethically controversial subject matter affecting the human body. It has even been said that an evolution of the patent regime into such areas should first have followed consultation of all stakeholders who would be affected and following social discussions about the desirability of the impact of life patents (Warner 2001).

2.5 A short history of biotechnology patents

However, it is worth pointing out that the patenting of life forms is not, in fact, historically a totally new phenomenon. Louis Pasteur, for example, patented a culture of yeast cells in 1873 (US Patent 141.072); adrenaline and insulin were patented in the early 20thcentury, in 1903 and 1923 respectively.[33] The US has had a *sui generis* plant patent system from as early as 1930[34] and shortly after the Second World War, Merck was granted patents on two products extracted from a microorganism, namely streptomycin and vitamin B12. In 1969, in Europe, the German Federal Supreme Court accepted the principle that animal breeding methods are patentable. However, the watershed was the famous US *Diamond versus Chakrabarty* case of 1980.

Diamond versus Chakrabarty involved a patent application, filed in 1972, for a human-made genetically engineered bacterium capable of breaking down crude oil. Because of this property, which is possessed by no naturally occurring bacteria, Chakrabarty's invention was believed to have significant value for the treatment of oil spills. The patent claims included process claims for a method of producing the bacteria, and these were granted from the outset, but what was first initially refused were the claims to the unique bacteria themselves. The US patent office had argued that microorganisms are products of nature and that living things are not patentable subject matter as contemplated under the US Constitution. The US Supreme Court, on the other hand, ruled by a narrow majority of 5 to 4 that a microorganism could be classed as a 'composition of matter 'or 'an article of manufacture' for patenting purposes and therefore be treated as a patentable invention. It stated that Congress had intended patentable subject matter to "include anything under the sun that is made by man." [35] This

[33] In 1724 Thomas Greening was granted an English patent for 'grafting or budding the English elm upon the stock of the Dutch elm'. In 1785, Philip Le Brocq acquired a patent for 'rearing, cultivating, training, and bringing to perfection, all kinds of fruit trees, shrubs, and plants; protecting their leaves, blossoms, flowers and fruits'.

[34] The 1930 Plant Patent Act extended patent protection only to certain asexually reproduced plants. In 1985 this was extended to sexually reproducing plants.

[35] 447 U.S. 303 (1980). Decision written by Chief Justice Warren E Burger.

decision upheld the earlier favourable Appeal ruling (subsequently challenged by Sydney Diamond, Commissioner of Patents and Trademarks) which had determined that, "*the fact that microorganisms are alive is without legal significance for purposes of the patent law.*"

The scope of patent protection in the US had now been officially extended by the courts, rather than the legislature, not only to microorganism *products* but to microorganisms themselves. [36] The Chakrabarty decision emboldened the biotechnology industry to argue that patents should issue on whole plants, on seeds, animals, tissues, cells, genes and other materials that had commercial value. In congressional hearings on the Biotechnology Competitiveness Act (which passed in the Senate in 1988), expert witnesses testified that the United States was locked in a "global race against time to assure our eminence in biotechnology"; a race in which the PTO had an important role to play (Sagoff 1998).

In the same year as the Supreme Court decision, an application from Stanford University filed in the early 1970's for a recombinant DNA sequence patent – based on what is found in nature but recombined from different sources through human intervention, was also granted.[37] This breakthrough in life sciences has had major impacts on biotechnology, medicine and research, the first licensed drug generated using recombinant DNA technology being that of recombinant human insulin that replaces animal derived substitutes (Johnson 1983).[38]

---

[36] Europe, however, had in fact for a number of years already accepted patents for micro-organisms. For example, in 1975, the German Federal Supreme Court ruled that micro-organisms were patentable. Also, the UK patent office had already granted a patent on Chakrabarty's oil eating bacterium during the 1970s without publicity. But it was only in 1988 that the EPO granted the first patent on plants.

[37] Recombinant DNA (rDNA) molecules are DNA molecules formed by laboratory methods of genetic recombination (such as molecular cloning) to bring together genetic material from multiple sources, creating sequences that would not otherwise be found in biological organisms. The patent was that of Hughes, S. (2001).

[38] Other highly useful and important applications of this type of technology include: recombinant chymosin- Chymosin, found in rennet, is an enzyme required to manufacture cheese. It was the first genetically engineered food additive used commercially. Also the following: recombinant human growth hormone (HGH, somatotropin) - before recombinant HGH became available, HGH for therapeutic use was obtained from pituitary glands of

In 1985 the US Patent and Trade Mark Office (USPTO) affirmed, in an appeals case, the patentability of plants, seeds and plant tissue cultures.[39] In 1987, it also announced that it would accept applications for multicellular organisms. One year later, in 1988, the first-ever animal patent was granted for a 'transgenic nonhuman mammal' containing an activated oncogene sequence (subsequently known as the Harvard OncoMouse). The activated oncogene significantly increases the mouse's susceptibility to cancer, and thus makes the mouse suitable for cancer research.[40]

2.6 In patent law living organisms 'are just chemical substances'

The proliferation, since the 1980s, of granted DNA sequence patents,[41] as well as other 'live' substance patents, including for plants and transgenic animals[42], has its technical explanation in the fact that for patent purposes these are treated as if they are *chemical substances* rather than natural products, which have been patentable for at least 150 years.[43] In the patent laws of Europe and North America an invention can be

---

cadavers; recombinant blood factor VIII -before the development of recombinant factor VIII, the protein was obtained by processing large quantities of human blood from multiple donors, which carried a very high risk of transmission of blood borne infectious diseases, for example HIV and hepatitis B; diagnosis of infection with HIV - each of the three widely used methods for diagnosing HIV infection has been developed using recombinant DNA; Golden Rice -a recombinant variety of rice that has been engineered to express the enzymes responsible for β-carotene biosynthesis. This variety of rice is said to hold substantial promise for reducing the incidence of vitamin A deficiency in the world's population. Herbicide and insect resistant corps have also been developed based on recombinant technology.

[39] Subsequently confirmed in December 2001 by the US Supreme Court.

[40] *US Patent no. 4,736,866*. The patent was also granted in Europe after a complex appeal: Board of Appeal of the European Patent Office, Decision of 6 July 2004, T 315/03.

[41] Another major type of DNA sequence which is currently patent eligible in many jurisdictions is complementary DNA (cDNA). This is a sequence of DNA as found in nature but without so-called 'junk' DNA or intron (intervening DNA sequences) which are not coding sequences – but whose role is not really understood (this is a type of sequencing allowed in Myriad – see below, fn. 56. Many critics, however, remain sceptical about the inventiveness of such patents and are vocal about the perceived undesirable restrictions placed on other scientists working within the same or similar fields.

[42] The granting of a patent on the OncoMouse caused fierce controversy in Europe and on-going opposition by civil society organizations.

[43] In 1995, the Opposition Division of the European Patent Office (EPO) declared DNA to be 'not "life", but a chemical substance which carries genetic information', and therefore patentable just as any other chemicals are.

claimed on something that occurs in nature as long as it is extracted and made available for industrial utilisation for the first time, and as long as there is a change resulting from human intervention , such as by adding something to it, for example a gene, subtracting something from it (by purifying it), mixing it with something else to create a new or synergistic effect, or structurally modifying it so that it differs in an identifiable manner from what it was before. The European Patent Office (EPO) guidelines for examination, for example, states that:

> *If a substance found in nature has first to be isolated from its surroundings and a process for obtaining it is developed, that process is patentable. Moreover, if the substance can be properly characterized either by its structure, by a process by which it is obtained by other parameters... and it is 'new' in the absolute sense of not having a previously recognised existence, than the substance per se may be patentable.*[44]

In the 2013 Manual for Patent examiners in the UK Intellectual Property Office, it is stated that "... *it is established practice that a natural substance which has been isolated for the first time and which had no previously recognised existence, does not lack novelty because it has always been present in nature*". Also, living matter claims are examined for inventiveness in a similar fashion to that which applies to chemical compounds. [45]

---

[44] *Guidelines for Examination in the European Patent Office*, Munich: EPO, at Part C – IV,2.3. The EU Biotechnology Directive Article 3(2) states: "Biological material which is isolated from its natural environment or produced by means of a technical process may be the subject of an invention even if it previously occurred in nature. We might note that a completely opposite approach is taken by the Andean Community's Common Intellectual Property Regime (Decision 486), which denied the possibility of inventiveness being claimed on any living thing, complete or partial, as existing in nature, even if separated, including the genome or germplasm of any living thing. In the recent *Myriad* case, the US Supreme Court held that "genes and the information they encode are not patent eligible under §101 simply because they have been isolated from the surrounding genetic material" (thereby reversing the until recent USPTO practice on protection of gene sequences as they occur in nature). However, this judgement made no statement on gene method patents, new applications of knowledge about genes or "the patentability of DNA in which the order of the naturally occurring nucleotides has been altered, [since] scientific alteration of the genetic code presents a different inquiry." [such as complementary or recombinant DNA] http://www.supremecourt.gov/opinions/12pdf/12-398_1b7d.pdf.

[45] Examination Guidelines for Patent Applications relating to Biotechnological Inventions in the Intellectual Property Office, par, 9 & 26. Intellectual Property Office.

2.7 DNA sequencing - controversies

A particularly controversial area of patenting has been that of DNA sequencing and gene mapping. The reason for the controversy has been mainly the issue of inventiveness – can information on how DNA is sequenced – i.e. finding out the order of the four base molecules or nucleotides in all of the chromosomes as found in nature, i.e. adenine (A), guanine (G), thymine (T) and cytosine (C), where most of the work is done mechanically by a computer, really be anything else but a discovery rather than an invention? Why should a private or even a public organisation, have a monopoly over this information, part of nature and common heritage, thereby also impeding the work of other researchers on the genomes of plants, animals and humans?

The fact is that many thousands of patents on DNA have been granted, mostly to biotech and pharmaceutical companies in anticipation of their future worth where genetic targets might be turned into drugs.[46] This of course has led to issues similar to other industries where patents are heavily used (such as electronics), with different aspects of the gene receiving individual patents owned by different companies or individuals. This leads to 'stacks' or 'thickets' making it difficult to know who owns what intellectual property, and acting as a deterrent to independent biomedical scientists from conducting research on particular segments of DNA. The US Patent Office was so overwhelmed by the number of applications at one stage – in particular when better sequencing techniques made results easier to gather – that by 1999 they had to modify the rules somewhat, requiring that in order to secure a patent any newly identified sequence had also to be accompanied by information on what proteins they coded, what the protein did and how to achieve that effect.

The most widely reported controversy regarding gene patenting involved the company *Myriad Genetics*. The company had obtained patents on genetic tests based on two identified breast cancer genes, BRCA1 and BRCA2, and proceeded to charge high

---

[46] It is estimated that around 50,000 such patents are owned by US companies, about 3-5000 being on human genes (about 20% of all human genes).

prices for its tests. [47]It also began enforcing its patent rights against other laboratories doing tests on these genes. Because potential breast cancer victims were being forced to pay thousands of dollars for a test – and other companies were being prevented from offering cheaper tests even based on their own research – a group of patients decided to challenge the validity of Myriad's patents in court. The case was finally decided by the Supreme Court of the USA on June 13, 2013 and resulted in the unanimous overturning of Myriad's claims. The Justices affirmed that the BRCA1 and BRCA2 genes constituted naturally occurring objects and as such were not patentable. However, the Court asserted that certain kinds of DNA molecules made in the lab and not occurring in nature, such as complementary DNA, would not fall under their ruling and hence could be patented. Although there are still unclarified questions about this ruling, one immediate effect has been that other companies have been able to offer genetic testing on these genes at a fraction of Myriad's price (Stevens 2016: 89); (Lopatto 2015).
Gene patents, despite the Myriad ruling, still have the potential of blocking research. Also, Myriad applied only to the US whereas gene patenting now occurs around the world. [48]

2.8 The Human Genome Project

There is potentially enormous scientific and potential social value in DNA sequencing, possibly leading to breakthroughs in medicine, agriculture, animal welfare and other areas. As we have seen with Myriad's discoveries, knowledge about a particular gene can have specific healthcare implications. Knowing what a defective gene looks like leads to genetic diagnostic tests and therapeutic or preventative interventions. If we could know the constitution of all of the 23 chromosome pairs of a human being – the whole human genome – this would be information of huge value to sci-

[47] The most famous person who used this test is the Hollywood actress Angelina Jolie, who decided to have a preventative double mastectomy on the basis of this test (Jolie, 2013).

[48] However, in Australia a similar challenge against Myriad resulted in a High Court decision in October 2015 to invalidate their patents.

entists worldwide. But because of the enormous complexity of the human genome (with billions of pieces of information) the US Government, through its National Institutes of Health, decided that the way forward would be to initiate a collaborative project – the **Human Genome Project** – on an international scale that included laboratories in the United Kingdom, France, Germany, Japan and China. It was a project that cost billions of dollars, which lasted many years (1990-2003).[49] Public funded organisations involved in the Human Genome Project early on mutually agreed not to patent their findings (where patent rules allowed it) but to immediately put it in the public domain.

Pope John Paul II, in an address to the Pontifical Academy of Sciences on 28 October 1998, also expressed his strong view that discoveries about the human genome should not become the property of a small group through patenting but be available to the whole scientific community.[50]

Although the project did not quite lead to the discoveries hoped for, and to some surprises (most of the sequenced DNA did not comprise genes and had unknown function; far fewer genes were actually discovered – around 20,000 – much less than the original 100,000 estimate and even far less than many other species), it kick-started many derivative projects and bio-industries.

The project opened up new panoramas for human welfare, but there were other aspects of this new information that continue to cause concern. Information showing genetic weaknesses in a person's genome could lead to the possibility of discrimination (for example health care insurance or employment opportunities), and to morally questionable programmes such as human enhancement, and morally objectionable programmes of selective discard-

[49] For a short account of the background, history and consequences of the Human Genome Project, see (Hood 2013); (Stevens 2016: 174-191).

[50] "... we rejoice that numerous researchers have refused to allow discoveries made about the genome to be patented. Since the human body is not an object that can be disposed of at will, the results of research should be made available to the whole scientific community and cannot be the property of a small group.". Address to the Plenary Session 'Human Genome; Alternative Energy Sources for Developing Countries; the Fundamental Principles of Mathematics; and Artificial Intelligence' 28 October 1998. Pontifical Academy of Sciences (1994).

ing of fertilised human embryos through preimplantation genetic diagnosis (PGD), eugenic selection of sperm from sperm banks for IVF, and human cloning.
As a consequence, the project, even before its full results were revealed, stimulated debate and comment around the world, including at the level of the UN at the International Bioethics committee of UNESCO, which lead to the Universal Declaration on the Human Genome and Human Rights (11 November 1998). It has also led to pronouncements from Church authorities as well as ample commentary from theologians and ethicists.[51]

2.9 Objections to genetic and life form patents - GM seeds and GM food

Earlier in this paper, when we looked at the use of Genetic Resources, the ethical questions that arose were concerned with social justice, the use of patents in a way that was fair to the custodians and originators of Genetic Resources and Traditional Knowledge. Other major ethical issues that arise from patenting life forms relate to the patenting of seeds that have been genetically engineered or modified specifically for better crop performance – GM seeds.[52] There is a vast literature on this subject – which includes especially food security and biodiversity in the developing world. The ethical issues are primarily those to do with social and distributive justice – the rights of farmers to reseed and to sell seed.[53] Advocacy

[51] UNESCO (1997). Article 2(a) states "Everyone has a right to respect for their dignity and for their rights regardless of their genetic characteristics". Article 4 states "The human genome in its natural state shall not give rise to financial gains" and Article 11: "Practices which are contrary to human dignity, such as reproductive cloning of human beings, shall not be permitted…". The report of the 4th Session of the International Bioethics committee can be found at See also Libreria Editrice Vaticana, Città del Vaticano (1999): 509. Also the papers published from the Pontifical Academy of Sciences, together with the plenary address of Pope John Paul II: www.casinapioiv.va A very useful resource is also the paper by Fleming.

[52] Seeds produced by professional breeders can also be protected in other ways – principally through plant breeders' rights such as UPOV. Other ways are through deliberately making seeds infertile by engineered 'killer' or 'terminator' genes, forcing a farmer to go back to the supplier each year. See the critical article on Monsanto in India: (Shiva 2013).

[53] See Biotechnology and society, Part IV Genetically Modified Foods: 97-129; and the references for further reading. In terms of advocacy groups, Greenpeace is one of the groups behind No patents on seeds [website]. On the home page is the 'missionstatement': *The*

groups such as Greenpeace are at the forefront of the attack on patents in relation to seeds which build unhealthy dependencies of farmers on suppliers. Companies such as Monsanto, who control around 23% of world seed supply and 87% of GM seeds, also sell agrochemicals (fertilizers and pesticides) against which their GM seeds are resistant. Farmers are often contractually obliged to purchase these goods along with the seeds.[54]

The main issue with GM seeds, however, is not so much control as fear of their environmental and health effects. This has resulted in the successful lobbying of NGOs such as Greenpeace such that Europe and most countries in Africa in particular do not allow this innovation to be introduced.

In line with the recommendations of the Pontifical Academy of Sciences on GM foods and a growing scientific consensus, Pope Francis recently in his encyclical *Laudato Si'* gave a cautious welcome, if used responsibly, of GM technology and foods (Pontifical Academy of Sciences 2009).[55] In a presentation at the Von Hügel

---

organisations behind No Patents On Seeds are especially concerned about increasing number of patents on plants, seeds and farm animals and their impact on farmers, breeders, innovation and biodiversity. These patents create new dependencies for farmers, breeders, food producers and consumers. These patents have to be regarded as misappropriation of basic resources in farm and food production and as general abuse of patent law. We call for an urgent re-think of European patent law in biotechnology and plant breeding and to support clear regulations that exclude from patentability processes for breeding, genetic material, plants and animals and food derived thereof. See also the Council of TRIPS document IP/C/W/369/Rev.1 Review of the Provisions of Article 27.2(b). Summary of Issues raised and points made.

[54] Monsanto's first generation GM products included potato, cotton and maize which were modified by the addition of a gene from the soil-dwelling bacterium *Bacillus thuringiensis* (Bt). Bt produces a toxin that kills insects. These crops were therefore more resistant to insects and required less pesticide. In another modification soybeans, canola and maize were given a gene from another microorganism, that made them resistant to the herbicide glyphosate, also developed by Monsanto, the main ingredient in the broad-spectrum herbicide known as 'Roundup'. Because this reduced the need for weeding crops could be planted closer together thereby increasing yield. Such crops are known as 'Roundup Ready'. Monsanto also developed seeds that contained both traits. For a list of genetically modified foods approved for use in the USA see International Service for the Acquisition of Agri-Biotech Applications [website].

[55] Cf. the UN Development Report 2001, United Nations (2001); Intervention of Prof Werner Arber, President of the PAS, XIII Ordinary General Assembly of the Synod of Bishops 7-28 October 2012, Holy See Press Office. Synodus Episcoporum Bulletin. See also Wüger – Cottier 2005.

Institute's international Patents on Life conference held 4-5 September 2015 in Cambridge, UK, Michael Kock, the global head of Intellectual Property at Syngenta, one of the top five global seed companies in the world, referred to this encyclical in forwarding his opinion that, given what we know today of the potential benefits of GM crops to alleviate hunger and disease, it could in fact be said to be morally irresponsible to prevent their use where needed.[56] He stated:

> *"Innovation and technology can be a blessing as much as a curse. Pope Francis in his recent encyclical letter "Laudato Si" recognizes that we are the beneficiaries of technological innovation including biotechnology: "[S]cience and technology are wonderful products of a God-given human creativity", which have "remedied countless evils which used to harm and limit human beings." With respect to genetic modification of plants, Pope Francis considers that "[t]he risks involved are not always due to the techniques used, but rather to their improper or excessive application. Genetic mutations, in fact, have often been, and continue to be, caused by nature itself." While "no conclusive proof exists that GM cereals may be harmful to human beings, and in some regions their use [...] has helped to resolve problems, there remain a number of significant difficulties [...]. In many places, following the introduction of these crops, productive land is concentrated in the hands of a few owners[...]"*
>
> *While technology is certainly not "neutral", in its purest sense it is primarily its purpose and use which makes it beneficial or problematic. Not using technology in view of the current challenges is ethically and morally questionable. ... It is easy for people in the developed world - where healthy and nutritious food is plentiful and affordable to reject new technologies. Certainly, the moratorium for the cultivation of genetically modified crops in Europe has a negative impact on the competitiveness of EU agriculture. The now necessary importation of millions of tons of genetically modified soybeans every year for the EU meat production industry make the position rather hypocritical and shifts the issue from Europe to other places in the world. Such attitudes become worse and ethically questionable if imposed with missionary zeal on the poor in developing countries. For example, NGO activism blocks the utilization of 'Golden Rice' to reduce blindness and other Vitamin A deficiency induced diseas-*

[56] Quotes from Pope Francis are from the section on new biological technologies of *Laudato Si'*, nn. 130-136. Michael Kock's paper will be published in full in 2018 together with other conference contributions by Cambridge University Press.

es, which are prevalent among the poor whose diets are based mainly on rice. Patrick Moore, a former Greenpeace director, today calls the NGO campaign "a crime against humanity" due to which 20 million people, mostly children, have died. The innovation behind Golden Rice has been recently recognized by Patents for Humanity Award of The White House Office of Science and Technology Policy and the U.S. Patent and Trademark Office. In conclusion, the use of technology and innovation can be a positive contribution to counter today's challenges, as long as this use is ethical and sustainable."

2.10 Morality and patent law

The following reflections deal with the question: can patents, which are a tool to promote innovation, be objectionable as immoral in themselves or only in their commercial use?

In the ground-breaking *Diamond vs. Chakrabarty* case, the US Supreme court was required to regard living organisms as simple material objects. Their changing nature, their ability to reproduce, and other factors proper to living organisms were, for the purposes of patent law determination, ignored. Following, the OncoMouse patent award, one activist wrote, "in one regulatory stroke, the Patent Office reduced the entire animal kingdom to the lowly status of a commercial commodity, indistinguishable from electric toasters and automobiles." (Rifkin 1989). Is this a true statement of affairs today regarding the whole modern biotechnology industry?

Some of the questions that present themselves with this development in patent law include:

a) Is a materialistic understanding of life being encouraged by the patent system, fostering a scientific ethos that disregards moral, ethical and spiritual aspects of life, to which religious organisations and the Catholic Church are required to respond?
b) Is unethical scientific research and experimentation being encouraged and endorsed by this 'neutral' or amoral approach to patenting?
c) Does the commercial exploitation of patented life forms necessarily reduce 'life' to a market commodity?

Such questions arise from a basic philosophical position that there are intrinsic values that require respecting, that draw on the conception of 'natural law' of what is proper and right and what is improper and wrong. This of course is part of the Christian tradition. Separately, but related to these questions, are others that follow from the premise that if patenting of life forms is acceptable in principle due to the developments in the biotechnology industries (viewing the patent system from its angle of providing social/economic benefits in any given society where a grant is awarded), then we might ask the following important questions:

- where are the cut off points for what is acceptable and what is not acceptable subject matter, and what are the criteria for acceptability?
- what are the cut off points for interventions on human beings which ought to automatically trigger immorality exclusions?
- where should patent examiners or the courts interpreting patent law be looking to in order to determine the unethical or immoral nature of patent subject matter?
- is patent law the best determinant of what is acceptable in biotechnology research and development?

These and other questions regarding the morality of biological interventions and their potential status as 'inventions' before the intellectual property system require serious attention by religious ethicists in order to help inform public policy from the perspectives unique to their traditions. There is already a significant body of secular academic literature devoted to exclusions from patentability of immoral inventions from which theologians may draw.[57]

[57] Recent studies include Liddell's "Immorality and patents: The exclusion of inventions contrary to *ordre public* and morality", in Lever, A. ed (2012): 140-171 (see also the bibliography on pages 145-6); Sterckx, S. and Cockbain, J. (2012); Van Overwalle's "Biotechnology and Patents Global Standards, European Approaches and National Accents" in Wüger, E.D. Cottier, T. (2008): 77-108; idem, "Legal and ethical aspects of bio-patenting. Critical analysis of the EU Biotechnology Directive", in Drahos 2005: 145-157.

2.11 A critique of some interventions by religious authorities

Whereas statements from religious authorities and ethicists on the intrinsic morality of the research presented for patent protection is not lacking, the views that have been expressed regarding the case to prevent a patent on living matter from being granted have been generally far from articulate.

In a 1989 statement, "*Consultation on Respect for Life and the Environment,*" signed *inter alia*by the National Council of the Churches of Christ in the USA (an ecumenical association of 37 Christian faith groups that includes Protestant and Orthodox Churches) there was a call for a moratorium on transgenic animal research where DNA from different species are used to produce chimeras[58]. The statement, however, asserted that such patented technology "portends fundamental changes in the public's perception of, and attitude towards animals, which would be regarded as human creations, inventions, and commodities, rather than God's creation and subjects of nature" and requests sustained public debate on such matters as "the current practice of combining human with nonhuman genetic material, unknown risks to human life, the probable suffering of the animals in question, provision for their humane care, the risk of adverse environmental impacts, and the possibility of deleterious economic and social effects on farmers and consumers worldwide."[59]

---

[58] Transgenics allow scientists to develop organisms that express a novel trait not normally found in the species; for example, a type of rice known as Golden Rice (as mentioned above) has elevated levels of vitamin A. Scientists have also developed sunflowers that are resistant to mildew and cotton that resists insect damage. Possible transgenic combinations can be broken down generally into three categories (here "animal" refers to nonhumans): 1) plant-animal-human combinations, 2) animal-animal combinations, 3) animal-human combinations. An example of a plant-animal-human transgenic combination would be one in which the DNA of mouse and human tumour fragments is inserted into tobacco DNA. The harvested plants contain a potential vaccine against non-Hodgkin's lymphoma. Animal-human transgenic combinations represent a booming aspect of biotechnology, for example commercial companies seek to derive therapeutic proteins, such as monoclonal antibodies, from the milk of transgenic cows, goats, rabbits, and mice and use them to administer drugs in treatment of rheumatoid arthritis, cancer, and other autoimmune disorders. Cf. MacDonald Glenn, L. (2013). The author also posits ethical questions, including proportionality of human benefit to animal suffering and the possibilities of social abuse.

[59] *Iibid, 134.* Also mentioned in Koehler 1996.

Our comment to this statement is that the real point being made was that a patent necessarily gives a seal of approval to unethical research. However, patents themselves do not make animals as such human creations. The language used in the statement is therefore confusing.

A few years later, in a statement entitled *Joint Appeal Against Human and Animal Patenting* signed in May 1995 by a coalition of around 185 religious leaders, including individual US Roman Catholic bishops (but not representing the US Bishops' Conference as such) the following was stated:

> "*We, the undersigned religious leaders, oppose the patenting of human and animal life forms. We are disturbed by the U.S. Patent Office's recent decision to patent body parts and several genetically engineered animals. We believe that humans and animals are creations of God, not humans, and as such should not be patented as human inventions.*"[60]

One immediate critical observation to be made on this statement is that, although patenting of life forms as such are objected to, significantly there is no mention of plant life or microbial life which are also 'creations of God'. But this is not the only problem with the statement.

Mark Sagoff from the Institute for Philosophy and Public Policy at the University of Maryland, a critic of the above Joint Statement for the manner in which complicated issues had been simplistically articulated, argues for a better understanding on the part of religious leaders of the nature of patents:

> "*Representatives of the biotech industry argue that their religious critics fail to understand the purpose of patent law. According to the industry view, patents create temporary legal monopolies to encourage useful advances in knowledge; they have no moral or theological implications....: A patent on a gene does not confer ownership of that gene to the patent holder. It only provides temporary legal protections against attempts by other parties to commercialize the patent holder's discovery or invention. Lisa Raines, vice president of the Genzyme Corporation, summed up*

[60] A seminal, albeit critical, work that analyses this statement and describes other contributions of various religious bodies and representatives to the debate on gene patenting, is Chapman, A.R. (1999): chapter four Cf. *UN Economic and Social Council Committee on economic, social and cultural rights*: E/C.12/2000/12: 54. See also Sagoff 1998.

> *the industry view regarding this intervention by representatives of the Churches and others: "The religious leaders don't understand perhaps what our goals are. Our goals are not to play God; they are to play doctor... The differences between the two groups are not irreconcilable. The religious leaders are not opposed to biotechnology, and the industry has no interest in being declared the Creator of life."* (Sagoff 1998)

It would certainly have helped the credibility of the religious leaders if they had had pressed the point that, as Sagoff further puts it,

> "*patent policy should maintain the distinction between discovery and invention, between what already exists in nature and what human ingenuity creates. [Therefore] the intricacies of nature . . . ought not to be patentable.* (Sagoff 1998)

In explaining the background to the Joint Appealone of the signatories, Bishop William Benedict Friend of the Roman Catholic Diocese of Alexandria-Shreveport, later issued an explanation where, after summing up the current state of Catholic teaching on gene patenting (primarily by referring to the teaching of Pope John Paul II) makes the rather curious statement:

> "*it is important for Catholics to distinguish that there are two different sets of issues to be addressed: the patenting of human genes distinct from the patenting of <u>animal</u> genes. This distinction flows from the nature and transcendental vocation of the human persons versus the nature of animals. It implies a whole different set of moral principles and theological teachings.*" (Friend 1995)

It is curious as what seems to be the underlying understanding is that patenting involves somehow a possession of the physical object of the patent right, in this case a gene, which of course is not the case. Excluding information and techniques is different from owning a piece of living property. From the scientific point of view, besides, DNA is common to all orders of the plant and animal kingdom. Nonetheless, the Bishop was attempting to affirm the care and moral duty that researchers should have in recognising the human subject in research as having an innate dignity proper only to humans.

Our own reflection is that life form patenting as such, to the extent that it involves morally and ethically acceptable experimentation on plants and animals, should not provoke any strong opposition based on Catholic social or moral teaching. How patent rights, based on life form or genetic interventions, are exercised by companies, is another issue which has been amply discussed in this paper.

2.12 Moral exclusions in patent law

It is important to point out that in TRIPS there is in fact an optional morality exclusion clause that excludes from patentability inventions whose commercial exploitation would be immoral or go against public order. [61] Although US patent law does not have this explicit morality exclusion, in practice the judiciary would interpret the law in this direction according to basic principles of a patent's social utility. In Europe, on the other hand, there are two sets of legislation that provide both specific morality exclusions and the general provision, as found in TRIPS, which reflects a certain in-built flexibility to exclude objectionable patents, including, and perhaps particularly, dealing with life or genetic patents. These are the EU Biotechnology Directive and Article 53 and Rule 28 of the European Patent Convention. Thus, for example, embryonic stem cell patents that involve the destruction of the human embryo are specifically excluded from patentability.[62]

[61] Art 27(2): Members may exclude from patentability inventions, the prevention within their territory of the commercial exploitation of which is necessary to protect ordre public or morality, including to protect human, animal or plant life or health or to avoid serious prejudice to the environment, provided that such exclusion is not made merely because the exploitation is prohibited by their law.

[62] Article 6 (2) of the Directive 98/44/EC of the European Parliament and of the Council of 6 July 1998 on the legal protection of biotechnological inventions excludes from patentability on the basis of being contrary to ordre public or morality a) processes for cloning human beings; (b) processes for modifying the germ line genetic identity of human beings; (c) uses of human embryos for industrial or commercial purposes; (d) processes for modifying the genetic identity of animals which are likely to cause them suffering without any substantial medical benefit to man or animal, and also animals resulting from such processes. These are illustrative rather than exhaustive (being excluded 'in particular'). The legislature was unwilling to commit to other types of technology e.g. human/animal recombinant technology, embryonic stem cell etc., although there is some case law on the matter, such as Oliver Brüstle v Greenpeace eV, Judgment of the Court (Grand Chamber) of 18 October 2011, Case C-34/10. The Court ruled that an

The banning of the patenting of embryonic stem cell patenting (although not the research as such) has led to considerable controversy given that research funding is often linked to patentability and there is among many groups of scientists the desire to experiment in this area.[63]
The ongoing challenge, however, is how to decide what is immoral. The judiciary are ill-equipped to make this judgement in the absence of guidance from moral authorities. In practice this might be a bio-ethics committee in which the influence of the Churches varies from nation to nation reflecting its degree of secularism, although the decision of the European Court of Justice in a recent case, as well as the provisions of the European Biotechnology Directive, are in line with Catholic moral teaching. At the European Patent Organisation recent case law has determined that the guide line for morality should be based on the principle that "the concept of morality is related to the belief that some behaviour is right and acceptable whereas other behaviour is wrong, this belief being founded on the totality of the accepted norms which are deeply rooted in a particular culture. For the purposes of the EPC, the culture in question is the culture inherent in European society and civilisation."[64]
The Catholic Church, through its moral teaching and social doctrine, has an important role in influencing these deeply rooted beliefs of European culture. It also has a role in providing moral guidance to those involved in granting biotechnology or genetic patents – examiners and judges – which indirectly influence the formation of European moral values or encourage industrial investment in immoral technologies such as those that destroy human embryos.[65]

---

invention is excluded from patentability where the implementation of the process requires either the prior destruction of human embryos or their prior use as base material. Article 53 EPC states European patents shall not be granted in respect of a) inventions the commercial exploitation of which would be contrary to "ordre public" or morality; such exploitation shall not be deemed to be so contrary merely because it is prohibited by law or regulation in some or all of the Contracting States. Rule 28 incorporates Article 6(2) of the Directive.

[63] For a quick overview see www.regimbeau.eu. See also a recent study commissioned by the EU: Plomer – Torremans 2009.

[64] Plant Cells/Plant Genetic Systems. T356/93 (OJ) EPO 1995, 545, 6.

[65] A 'designer baby' patent in the US has recently been awarded that both reflects and will influence the acceptability of a new form of eugenics. Patent #8,543,339 grants

Conclusion: acceptance of genetic research and patenting by the Church

The Church's position on life form and genetic research and responsible patenting flows from accepting the role of human creative intervention in nature as a good, but only when it is simultaneously coupled with responsibility. The *Compendium of the Social Doctrine of the Church* states:

> "*nature is not a sacred or divine reality that man must leave alone. Rather, it is a gift offered by the Creator to the human community, entrusted to the intelligence and moral responsibility of men and women. For this reason the human person does not commit an illicit act when, out of respect for the order, beauty and usefulness of individual living beings and their function in the ecosystem, he intervenes by modifying some of their characteristics or properties.*" (*Compendium*: 473)

I would like to end this paper with some words of Pope John Paul II which are little known but which show his full support for genetic science (and implicitly for responsible patent protection). These are words directed to members of the Pontifical Academy of Sciences, composed of eminent scientists of all religious persuasions or none.

> "*Today, far from shutting herself up in an apologetic or defensive perspective, the Church rather makes herself the advocate of science, of reason, and of the freedom of research.*"
>
> "*I have no reason to be apprehensive for those experiments in biology that are performed by scientists who, like you, have a profound respect for the human person, since I am sure that they will contribute to the integral well-being of man. On the other hand, I condemn, in the most explicit and formal way, experimental manipulations of the human embryo, since the human being, from conception to death, cannot be exploited for any purpose whatsoever. Indeed, as the Second Vatican Council teaches, man is 'the only creature on earth God willed for itself'* (*Gaudium et Spes*, n. 24)

23andMe, a genetics testing company that sells at-home DNA kits directly to consumers, exclusive rights to genetic and computer technologies that would enable prospective parents to handpick a sperm or egg donor with whom they would be likely to produce a child born with certain traits that they desire.

"I have learned with satisfaction that among the subjects discussed during your study week you have focused attention on in vitro experiments which have yielded results for the cure of diseases related to chromosome defects. It is also to be hoped, with reference to your activities, that the new technique of modification of the genetic code, in particular cases of genetic or chromosomic diseases, will be a motive of hope for the great number of people affected by those maladies. It can also be thought that, through the transfer of genes, certain specific diseases can be cured, such as sickle-cell anaemia, which in many countries affects individuals of the same ethnic origin. It should likewise be recalled that some hereditary diseases can be avoided through progress in biological experimentation. The research of modern biology gives hope that the transfer and mutations of genes can ameliorate the condition of those who are affected by chromosomic diseases; in this way the smallest and weakest of human beings can be cured during their intrauterine life or in the period immediately after birth. Finally, I wish to recall, along with the few cases which I have cited that benefit from biological experimentation, the important advantages that come from the increase of food products and from the formation of new edible plant species for the benefit of all, especially people most in need."[66]

This author is of the view that, when done in a socially responsible way, and respecting the integrity and dignity of man, intellectual property and patents in particular can have an important role in promoting the work of genetic research, experimentation and diagnostics as blessed by Saint John Paul II.

[66] Pontifical Academy of Science, Address to the Study Week on the subject 'Modern Biological Experimentation', 23 October 1982, 4-6.

References

ABBOTT R., *Documenting Traditional Medical Knowledge, WIPO*, 2014. At: www.wipo.int/export/sites/www/tk/en/resources/pdf/medical_tk.pdf. Accessed 2017.

BRAITHWAITE – DRAHOS, *Global Business Regulation*, Cambridge, 2000.

CAREY, «The Social Mortgage of Intellectual Property» in: *Christian Social Thought Series* 11, 2007.

CARITAS IN VERITATE FOUNDATION, *Patents on Genetic resources? A Catholic Perspective for the World Intellectual Property Organization*: 76-108, 2013. At: www.fciv.org/downloads/FCIV%20WP3%20Patents%20on%20Genetic%20Resources.pdf. Accessed 2017.

CHAPMAN A.R., *Unprecedented Choices: Religious Ethics at the Frontiers of Genetic Science*, Fortress Press, 1999.

CHEN J., *There's No Such Thing as Biopiracy... and It's a Good Thing Too*, 2006. At: https://www.researchgate.net/publication/228191099. Accessed 2017.

CONVENTION ON BIOLOGICAL DIVERSITY, Article 2. At: www.cbd.int/convention/articles/default.shtml?a=cbd-02. Accessed 2017.

DEERE C., *The Implementation Game. The TRIPS Agreement and the Global Politics of Intellectual Property Reform in Developing Countries*: 8-9, Oxford, 2009.

ETC Group, At: http://www.etcgroup.org/issues/patents-biopiracy. Accessed 2017.

EUROPEAN COMMISSION, *Article 6 (2) of the Directive 98/44/EC of the European Parliament and of the Council of 6 July 1998 on the legal protection of biotechnological inventions*, 1998. At: http://eur-lex.europa.eu/legal-content/EN/TXT/?uri=celex-%3A31998L0044. Accessed 2017.

EUROPEAN PATENT OFFICE, *Guidelines for Examination in the European Patent Office*. At: http://www.supremecourt.gov/opinions/12pdf/12-398_1b7d.pdf . Accessed 2017

FLEMING J.I., *The Catholic Moral Tradition and The Genome Project and The Diversity Project*. At: www.bioethics.org.au/Resources/Online%20Articles/Other%20Articles/The%20Catholic%20moral%20tradition%20and%20the%20genome%20project.pdf . Accessed 2017

FRIEND W.B., *An Introduction to Genes and Patents*, 1995 . At: http://profiles.nlm.nih.gov/ps/access/DJBBMY.pdf. Accessed 2017.

PLOMER A. – TORREMANS P. eds., *Embryonic Stem Cell Patents: European Law and Ethics,* Oxford, 2009.

HO C. M., «Biopiracy and Beyond: A Consideration of Socio-Cultural Conflicts with Global Patent Policies.» *U. Mich. J. L. Reform* 433, 2006.

HOOD L., *The Human Genome Project: big science transforms biology and medicine*, 2013. Available at: https://genomemedicine.biomedcentral.com/articles/10.1186/gm483. Accessed 2017.

HUGHES S., «Making dollars out of DNA. The first major patent in biotechnology and the commercialization of molecular biology, 1974-1980». Ibiden *Isis; an international review devoted to the history of science and its cultural influences* 92 (3): 541–575, 2001.

INTELLECTUAL PROPERTY OFFICE, *Examination Guidelines for Patent Applications relating to Biotechnological Inventions in the Intellectual Property Office*. Accessible at: www.ipo.gov.uk/biotech.pdf. Accessed 2017

INTERNATIONAL SERVICE FOR THE ACQUISITION OF AGRI-BIOTECH APPLICATIONS, *GM Approval Database*. Available at www.isaaa.org/gmapprovaldatabase/default.asp. Accessed 2017.

JOHNSON I. S., «Human insulin from recombinant DNA technology». *Science* 219 (4585): 632–637, 1983.

JOLIE A., *My medical choice*. New York Times. Available 2013. At: www.nytimes.com/2013/05/14/opinion/my-medical-choice.html . Accessed 2017.

Journal of Markets and Morality, January 2012. At: www.mayoresearch.org/files/SOCIAL%20MORTGAGEjan302012.pdf. Accessed 2017.

KOEHLER G.A., *Bioindustry: A Description of California's Bioindustry and Summary of the Public Issues Affecting Its Development*, 1996. At www.library.ca.gov/CRB/96/CRB-96-007.pdf. Accessed 2017.

KOCK M., (2018). [Will be available in 2018 together with other conference contributions. To be published by C.U.P.]

LEE R., *Malaria Cure Based On Traditional Chinese Medicine Wins Tu Youyou Nobel Prize*, 2015. At: www.techtimes.com/articles/93754/20151012. Accessed 2017.

LEVER A. ed., *New Frontiers in the Philosophy of Intellectual Property*, Cambridge, 2012.

LIBRERIA EDITRICE VATICANA, *Human genome, human person and the society of the future. Proceedings of IV Assembly of the Pontifical Academy for Life* (Vaticano, 23-25 Feb 98), Città del Vaticano, 1999, pp. 509.

LIBRERIA EDITRICE VATICANA, *Compendium of the Social Doctrine of the Church*, 2004, 283. At http://www.vatican.va/roman_curia/pontifical_councils/justpeace/documents/rc_pc_justpeace_doc_20060526_compendio-dott-soc_en.html. Accessed 2017

LOPATTO E., *Genetic testing for breast cancer gets more affordable*, 2015. At: www.theverge.com/2015/4/21/8458553/color-breast-cancer-gene-testing-brca-myriad. Accessed 2017

MACDONALD GLENN L., *Ethical Issues in Genetic Engineering and Transgenics*, 2013. At: www.actionbioscience.org/biotechnology/glenn.html. Accessed 2017

MCDOUGALL P., *The cost and time involved in the discovery, development and authorisation of a new plant biotechnology derived trait*, 2011. At: https://croplife.org/wp-content/uploads/2014/04/Getting-a-Biotech-Crop-to-Market-Phillips-McDougall-Study.pdf . Accessed 2017.

MULLIN R., *Cost to develop new pharmaceutical drug now exceeds $2.5b*, 2014. At: https://www.scientificamerican.com/article/cost-to-develop-new-pharmaceutical-drug-now-exceeds-2-5b/. Accessed 2017.

NAGAN W.P., et al. *Misappropriation of Shuar Traditional Knowledge (TK) and Trade Secrets: A Case Study on Biopiracy in the Amazon*, 2010. At http://scholarship.law.ufl.edu/cgi/viewcontent.cgi?article=1345&;context=facultypub. Accessed 2017.

No Patents on Seeds, At: http://no-patents-on-seeds.org/. Accessed 2017.

O'BOYLE E.J., «Social Mortgage: Origins, Questions, Norms». In: *Journal of Markets and Morality*, January 2012. At: http://www.mayoresearch.org/files/SOCIAL%20MORTGAGE-jan302012.pdf. Accessed 2017.

PONTIFICAL ACADEMY FOR LIFE, At: www.academiavita.org. Accessed 2017.

PONTIFICAL ACADEMY OF SCIENCES, *Address to the Study Week on the subject Modern Biological Experimentation*, 23 October 1982.

PONTIFICAL ACADEMY OF SCIENCES, *Plenary Session of the Pontifical Academy of Sciences*, 1994. At: www.casinapioiv.va/content/dam/accademia/pdf/sv92pas.pdf. Accessed 2017.

PONTIFICAL ACADEMY OF SCIENCES, *Address to the Plenary Session 'Human Genome; Alternative Energy Sources for Developing Countries; the Fundamental Principles of Mathematics; and Artificial Intelligence*, 1998. At: www.casinapioiv.va/content/accademia/en/magisterium/johnpaulii/28october1994.html.

PONTIFICAL ACADEMY OF SCIENCES, *Transgenic Plants for Food Security in the Context of Development*, 2009. At: www.casinapioiv.va/content/accademia/en/publications/scriptavaria/transgenic.html.

PONTIFICAL COUNCIL FOR JUSTICE AND PEACE, *Compendium of the Social Doctrine of the Church*, 2005. At: www.vatican.va/roman_curia/pontifical_councils/justpeace/documents/rc_pc_justpeace_doc_20060526_compendio-dott-soc_en.html. Accessed 2017.

RIFKIN J., «Patenting of Animals – Ethical considerations». In: *New Developments in Biotechnology: Patenting Life – Special Report*, April 1989. At: www.princeton.edu/~ota/disk1/1989/8924/8924.PDF. Accessed 2017.

ROBINSON D.F., *Confronting Biopiracy: Challenges, Cases and International Debates*, London, Earthscan, 2010.

SAGOFF M., «Patented Genes: An Ethical Appraisal». In: *Science and Technology Online: Perspectives*, Spring 1998. Available at: http://online.sfsu.edu/rone/GEessays/PatentedGeneEthics.htm. Accessed 2017.

SCIENCE MUSEUM, Louis Pasteur (1822-95). At: www.acs.org/content/acs/en/education/whatischemistry/landmarks/flemingpenicillin. Accessed 2017.

SHIVA V., *The Seeds of Suicide: How Monsanto Destroys Farming*, 2013. At: http://www.globalresearch.ca/the-seeds-of-suicide-how-monsanto-destroys-farming/5329947. Accessed 2017

STEVENS H., *Biotechnology and Society. An Introduction*, Chicago, 2016.

SYNOD OF BISHOPS, Intervention of Prof Werner Arber, President of the PAS, XIII Ordinary General Assembly of the Synod of Bishops 7-28 October 2012, Holy See Press Office, 2012.

TZEN WONG E. – DUTFIELD G., *Intellectual Property and Human Development, Current Trends and Future Developments*, Cambridge, 2011.

UNITED NATIONS ECONOMIC AND SOCIAL COUNCIL COMMITTEE ON ECONOMIC, SOCIAL AND CULTURAL RIGHTS, 2000. E/C.12/2000/12, p. 54.

UNITED NATIONS, UN General Assembly interim report of the Special Rapporteur on the right to food, 2001. At www.farmersrights.org/pdf/RighttoFood-N0942473.pdf. Accessed 2017.

UNITED NATIONS, *Intellectual Property in the World Trade Organization: Turning it into developing countries' real property*, 2010. At http://unctad.org/en/Docs/ditctncd20068_en.pdf. Accessed 2017.

UNITED NATIONS EDUCATIONAL, SCIENTIFIC AND CULTURAL ORGANIZATION, *Report of the 4th Session of the International Bioethics committee*. At: http://portal.unesco.org/shs/es/files/2310/10596570051 4VolumeI_en.pdf/4VolumeI_en.pdf.

UNESCO, Universal Declaration on the Human Genome and Human Rights, 1997. At http://portal.unesco.org/en/ev.php-URL_ID=13177&URL_DO=DO_TOPIC&URL_SECTION=201.html. Accessed 2017.

VAN OVERWALLE G., «Biotechnology and Patents: Global Standards, European Approaches and National Accents». *Genetic Engineering and the World Trade System*: 77-108, 2008.

VARIOUS, *Patents on Life: Through the Lenses of Law, Religious Faith and Social Justice*, 2015. At: www.vhi.st-edmunds.cam.ac.uk/events/past-events/conferences/patents-conference-2015. Accessed 2017.

VON HÜGEL INSTITUTE FOR CRITICAL CATHOLIC INQUIRY, *International Conference, Patents on Life*, 2015. At: http://www.vhi.st-edmunds.cam.ac.uk/events/past-events/conferences/patents-conference-2015. Accessed 2017.

WARNER K.D., «Are Life Patents Ethical? Conflict between Catholic Social Teaching and Agricultural Biotechnology's patent regime» in: *Journal of Agricultural and Environmental Ethics*, 301-319, 2001.

WORLD HEALTH ORGANIZATION, Available at: www.who.int/globalchange/ecosystems/biodiversity/en/. Accessed 2017.

WORLD INTELLECTUAL PROPERTY ORGANIZATION, *Holysee on intellectual property and genetic resources, traditional knowledge and folklore*, 2001. At: http://www.wipo.int/edocs/mdocs/tk/en/wipo_grtkf_ic_1/wipo_grtkf_ic_1_7.pdf. Accessed 2017.

DRAHOS P. ed., *Death of Patents*, Wotney, Lawtext Publishing, 2005.

WÜGER E.D. – COTTIER T., *Genetic Engineering and the World Trade System*, Cambridge, 2008.

Lecture n. 3

SCIENCE
The Cyborg and Cyborgization

Dr. Paolo Benanti Tor
Faculty of Theology
Pontifical Gregorian University

1. What is a Cyborg?

In the early 1960s, Manfred E. Clynes and Nathan S. Kline, two NASA researchers, coined the term cyborg from the combination of the words cybernetic and organism (Gray, Figueroa-Sarriera and Mentor 1995). Clynes and Kline wrote: "for the exogenously extended organizational complex functioning as an integrated homeostatic system unconsciously, we propose the term 'cyborg.' The cyborg deliberately controls function of the organism in order to adapt it to new environments" (Clynes – Kline 1995). The proposed "purpose" of the cyborg was to provide a system in which robot-like behaviors and actions are integrated into an automatic and allostatically balanced organization that is in some way integrated into human function, leaving humans free to explore and to flourish.
In an interview, Clynes reflected upon the term: "…I thought it would be good to have a new concept of a person who can free themselves from the constraints of the environment to the extent that they wished. And I coined this word cyborg." (Gray 1995) The idea was to liberate man to explore and discover space. It seemed necessary to afford the bodily freedom to exist in another part of the universe without the limitations imposed by having evolved on Earth. Clynes posed that the cyborg would enable humans to not change nature but to simply make use of biological faculties, without having to waste energies on adjusting the living function necessary for the maintenance of life (Gray 1995).
Clyne's original idea of cyborg was popularized in science fiction (Sci-Fi) by comics, and movies, and has become a symbol of the 20th century's complex relationship between man and technology

(Caronia 1985). Technological developments and certain influences of culture have allowed the original concept of the cyborg to acquire new scientific and socio-cultural significance over the last decades. Indeed, human-machine interface can access at a variety of levels, and Clynes and Kline originally classified those iterations as four possible tiers of cyborgization.

Cyborg I is a human with prosthetic organs, but these alterations do not change human nature. An artificial organ or limb, for example, functions to maintain the internal environment, or articulate in external environments, and does not affect basic human nature to any extent. Thus Cyborg I was concerned with physiologic enhancement (Hables 1995).

Cyborg II is an "evolution" of cyborg that entails both physiological and emotional aspects of human function (Hables 1995). Cyborg II shifts the focus away from the view of human and machine systems working interdependently, to an integrative network whose parts are interconnected and linked to physical and emotional well-being.

Cyborg III makes use of the abilities to both alter the products of genes, and also insert new elements into the existing genome. Cyborg III permits the ability to design new geno- and phenotypes, making use of knowledge to decrypt ways to improve human nature, make humans less destructive and more creative, and enable flourishing via the heightened enjoyment of the various satisfactions of life (Hables 1995). This represents a profound distinction in the construct of a cybernetic organism: the cyborg is no longer a product for the conquest of space, but is now a contemporary paradigm to facilitate happiness and humans flourishing.

Cyborg IV represents the transgenerational fusion of humans and technology, such that the integrated changes will be transmitted to and propagate in subsequent generations. Indeed, current research in epigenetic and even genomic modification through biotechnologic intervention and integration may be initial – yet tangibly important steps toward such ends. Clearly this represents participatory evolution that will really change human beings and human nature, and we must acknowledge this capacity to be for better or for worse (Hables 1995).

Cyborg V is the ultimate articulation of the machine-human interface. Eventually, the brain-mind may be linked to devices and media, and may thus not need the body for existence. Cyborg V would allow the essence of consciousness to survive as removed from the vicissitudes of the body, with expanded functionality, more highly evolved feeling, with further developed empathy (Hables 1995).

2. Posthumanism and the cyborg: Human enhancement

The concept of the cyborg encounters new anthropological perspectives in light of contemporary trans- and post-human ideas. The basic model of humanism, precisely because it (positively) appeals to the notion of a core humanity or common essential feature in term of which human beings can be defined and understood, continues to enjoy 'common sense' status of in Western culture (Badmington 2000). Yet, recent discourse focusing upon the ability of change the human, and the evolution nature of humans as a species in transition, such fixed concepts of the human and human nature as restrictive.
Thus, the cyborg is seen as an icon of the crisis in humanism. Although the subject of considerable debate, there is building literature claiming that the reign of humanity is at a critical if not questionable juncture. Such uncertainty about human futures is the face of neuro, nano, and genotechnology. The explicit question remains what humanity might, or will "become". These questions are part of the movement of trans-humans (Bostrom 2005) and post-humanism (Badmington 2000), that views the human as a collection of heterogeneous components, a material-informational entity whose boundaries undergo continuous construction and reconstruction (Hayles 1999). In this way, there is no "given" or pre-established autonomous self, because the self can only be made. Even the subjectivity of technologically unaltered individuals is perceived as an artifact (Waters 2006). Thus, information patterns are more important – if not – essential to the nature of being than any material instantiation. Embodiment in a biological substrate is seen as an accident of history rather than an inevitability of life (Hook 2004).

Based on the assumption that the universe consists principally of information, any entity that can code process, and sense data could be considered "alive" (Waters 2006).
In this context, the cyborg can help humanity to achieve the ancient dream of everlasting life: if humans can become pure information then they can also achieve immortality. But the first problem that we must overcome is that information is always embedded in a medium and must be extracted. To become post-human (in this context) requires the ability to remove information from the body, and place it in a superior stratum (Waters 2006). Post-humanism disavows essential differences or absolute demarcations between bodily existence and generated computer domains, biological organisms and cybernetic mechanisms, and human goals and robot technology (Hayles 1999). This view leaves humanity facing a horizon of choice that may have no telos. Indeed, such ends may represent a "telos of no telos" that offers modernity a way out of its dilemma: if there is no "ultimate" fate, then suitable alternative(s) can be fabricated. Moreover, those constructing such alternative fates are simultaneously reconstructing themselves.
There is no need to wait for some Nietzschean vision of the Übermensch to evolve; such as being can be engineered. Thus, the future will be largely what we make it; what it is willed to be. The "telos of no telos" is in fact a *telos* of *techne* (Waters 2006). Such that *techne* (viz – the embrace of technology and knowledge and capabilities it affords) enables a telos that allows us to recognize that the world's underlying and universal feature is information. Since information has no inherent meaning it can be recast, conveyed and interpreted in virtually endless ways.
The fluidity of information means that all borders are temporary and any definition permissible (Waters 2006). Post-humanism implies that we must regard the inextricability of the co-evolution of science and society, and technology and society (Achterhuis 2001). Technological development, in short, is not an independent force impinging upon society from the outside, rather, it is a social activity in itself, which reflects the particular settings of time, place, dreams and purposes, and the relation between people (Noble 1978). Cyborgs are constructions lacking a fixed, pre-ex-

isting, natural identity; they are made, not born; they also partly reflect the predicament of contemporary human beings, whose existence is ineluctably bound to, and within a technological framework, and whose life-world has become dominated by technology (Munnik 2001).
Traditional philosophical anthropology is generally conceived as anthropo-ontology; it reflects the ways of "being human." But by the end of twentieth century these ways of being have become inseperably involved with technolog, and thuss\ anthropo-ontology becomes cyborg-ontology (Munnik 2001). This cyborg-ontology may have a subversive effect on society, because a cyborg can conceive itself only as a construction, it can never conceive itself as a moment in a history that involves nostalgic sentiments. A cyborg is radically excluded from these ideas and is untroubled by the fact. It is a being of everlasting dispersion; it has forever lost its innocence. For Donna Haraway, this is the subversive character of the cyborg (Munnik 2001). Within Bostrom's trans-humanist orientation, the intellectual and cultural movement that affirms the possibility and desirability of cyborgs to fundamentally improve the human condition through applied reason: by using technology to eliminate aging and enhance human intellectual, physical and emotional capacities (Bostrom 2003). The Singularity is the most extreme aspect of the transhumanist/post-humanist vision, an idea that humankind should engineer the next phase of its own evolution, and that human beings should be augmented and altered, even to the point of losing "humanity" (Hook 2004).

3. The key: complexity and emergence

Such visions of human enhancement espouse the cyborg concept and have gained momentum by relating to constructs of complexity. Complexity science is an interdisciplinary field that seeks to explain how large numbers of relatively simple entities organize themselves, without the involvement of a central controller, into a collective whole that creates patterns, uses information, and, in some cases, evolves and learns (Mitchell 2009, 349). A complex system is one that exhibits non-trivial emergent and self-organizing behaviors, for example: ant colonies, immune systems, and brains (Mitchell: 2009, 349).

Complexity and emergence become tools to enable a technological approach to biological systems, because many, if not all, complex biological systems have a fine-grained architecture that consist of large numbers of relatively simple elements that work together in a highly parallel fashion (Mitchell: 2009, 349). Complexity and emergence also are key to research on the cyborg due to the impact that information plays in the scientific landscape. Information and computation have increasingly assumed an ontological status equal to that of mass and energy - namely, as primitive components of reality. In biology in particular, the description of living systems as information processing networks has become common (Mitchell: 2009, 349), (Holland: 1999, 258). In biological systems information is not, as in a traditional computer, precisely or statically located in any locus, instead, it takes the form of dynamic patterns with the system's components to impart complexity (Mitchell: 2009, 349).

Viewing the human as a complex system opens new technological opportunities to achieve the cyborg: we can obtain enhancement through interaction with the simple elements that make up the body. Actually, if we look at a small part of the organism, this may well be heading toward some local state of equilibrium. On the cellular scale, the process of life depends upon a continuous flow of energy and entropy. But if we look at a single macromolecule or a group of macromolecules, then we may define a local region in which structures come into existence by self-assembly and which is heading toward equilibrium (Jones: 2008, 228). The self-assembly process resulting in the folding of protein is very attractive for its potential uses in human enhancement. Indeed manipulating the dynamics of protein folding would allow the design and construction of human-made machines on a nanoscale (Offenhäusser and Rinaldi 2009).

Development of technologies at the nanoscale may be key to cyborgization via biological enhancement through the integration of self-organizing bio-technologies. This approach, engaging complexity and emergence, may allow an iterative presentation of the ultimate level of the cyborg: a human-machine complex system that exhibits non-trivial emergent and self-organizing behaviors, or properties that are inherent to, and inextricable from the integrated bio-machine organism.

4. A project that overcomes complexity: NBIC

The National Science Foundation (NSF) report, Converging Technologies for Improving Human Performance. Nanotechnology, Biotechnology, Information Technology and Cognitive Science, designed a path to realize technology with which to achieve human enhancement (Roco and Bainbridge 2002). NSF identified four areas of development to achieve such goals and ends: Nanotechnology, Biotechnology, Information Technology and Cognitive Science (abbreviated NBIC), recognizing that these areas are currently in convergence, and that the power of this model lies in – and across from – such integrative convergence.

A fundamental aim of the NSF project was to implement effective governance for the best, and most rapid, development of NBIC technologies. According to the report at least five reasons compel work on such a project: 1. it is essential to identify new technologies that have the potential for improving human performance, especially those that are unlikely to be developed as a natural consequence of the day-to-day activities of single governmental, industrial or educational institutions; 2. government has an important role in setting long-term priorities that insure that an environment exists in which beneficial innovations will be developed; 3. while American science and technology clearly benefit the entire world, it is vital to recognize that technological superiority is important to the economic prosperity and national security of the United States; 4. NBIC is critical to the mission of various government agencies; and 5. science must offer society new visions of what is possible to achieve, and scientists must become stewards of such visions and possibilities (Bond et al. 2002)

The implementation of the NBIC model is possible due to four key elements. First, the convergence of diverse technologies is based upon material unity at the nanoscale and upon technology integration derived from that scale (Roco and Bainbridge 2002). To re-iterate, it may soon be possible to achieve human-technology integration by engaging nanoscale construction and control to assemble complex microscale elements, such as neurons and some elements of the immune system, and from this, link systems such as organs and/or organism functions to larger meso and macro-scale components. Progress in these domains is revolutionizing the entire landscape of technology (Bhushan: 2007, 1916). Second, is the realization of NBIC transforming tools including scientific instruments, analytical methodologies, and radically new materials systems (Roco and Bainbridge 2002). A project such as the enhancement of the human being can not achieved with a single technology, even if extremely advanced, but must involve the development of a set of tools to manipulate both technologies and biological elements. A third key element for achieving the NBIC model is the development of system approaches, mathematics, and computation in areas that allow further understanding and control of complex, hierarchical systems (Roco and Bainbridge 2002). Fourth, is that the potential uniquity of the present time in relation to the opportunities, conferred by the fusion of cutting-edge nano-neuro and cyberscientific and technologic advancements (Roco and Bainbridge 2002). These next levels of human enhancement will require a particular type of relationship between different NBIC technologies. The nature of this relationship has been described as:

> "…if the Cognitive Scientists can think it,
> the Nano people can build it,
> the Bio people can implement it,
> and the IT people can monitor and control it -
> we will achieve human enhancement" (Roco – Bainbridge 2002).

5. From cyborg technologies to frank cyborgization

The NBIC model of allows a mappable route toward developing those technologies necessary for such high level human enhancement. These can be considered to be cyborg technologies. It seems appropriate to start from this consideration as a basis to analyze the characteristics of the relationship between humans and technology. To be sure, cyborg technologies enable human-machine interfacing. The term interface we often refer to a computational model to define a connection device that enables communication between two components of a system with different operating codes. Thus, an interface is a boundary between two systems, and through it, it is possible to communicate information. Usually there is a distinction according to types of interface. Hardware is an interface that includes a number of devices that allow the physical transmission of information. This transmission can happen in different ways: mechanical, electronic or chemical. Software is an interface that works in a form of mediation, such as a language or a code. This interface includes a source of information, broadcaster, and a destination, i.e. a receiver. While hardware interfaces focus upon the mode of exchange and communication possibility, software interfaces analyze what (i.e.: quality of message) can be communicated for specific purposes. An interface in cyborg technologies defines a type of relationship between human and machine that obtains and transmits complexity and emergence: such an interface enables understanding reality as consisting of information; therefore, controlling the flow of information enables control of reality.

The interface of NBIC technology occurs through various levels of complexity and emergence. At a first level, interfacing consists of man-made nanomachines that can engage and/or modify with proteins (e.g. bio-nanodevices). This type of interface can entail electrochemical and mechanical relationships between the various components: a combination of these two types of relationship forms the basis of cyborg technologies. NBIC technologies treat cognitive phenomena as information through nano- micro and macroscale elements that constitute the complex systems of the human organism "hardware."

A second type of interfacing is the "soft interface." Soft interfacing is based upon hard interface to obtain a hierarchically higher level of complexity. Thus, while hard interfacing involves exchanging – and processing – information systems; soft interfacing refers to the cognitive processes and properties that such information gen-

erates or evokes. Cyborg technologies are hard and soft interfaces that work at different levels of complexity. In this way, we see the cyborg as an enhanced human that is linked to a combination of hard and soft interface technologies that enable and augment information acquisition, integration, and cognitive capacity.
The totality of the NBIC paradigm may produce a complex human-machine system. Although we are not yet able to create NBIC technologies that will perform at this level, we have technologies that interact with human users in physical, cognitive and emotional ways. So-called sociable technologies affect the people who use them, influence psychological development, human relationships, and additionally, they enter into the cognitive confusion, to prompt such questions as the nature of the mind and self (Turkle 2002). At present we are slowly, yet steadily realizing the cyborg. We can define this intermediate state "cyborgization" to be the social use of technologies to change how we think, act, and live. Such cyborgization prepares society to accept the idea of machine-based, -derived and -dependent human capabilities and enhancement. NSF identified five areas in which the interaction of the NBIC sciences can expand the cognitive and communicative aspects of human performance: 1) Human Cognome Project, 2) Personal Sensory Device Interfacing, 3) Enriched Communities, 4) Learning how to learn, and 5) Enhanced Tools for Creativity (Bainbridge et al. 2002). Each have potential to both change our vision of humanity and its limits, and affect means to such change. In sum, the cyborg is a term and entity whose content has been gradually changing as a function and consequence of technological and cultural contexts in which it is developed and utilized. Having established the current cultural landscape (i.e.: trans – and post-humanism) and described its technological environment (viz - the NBIC paradigm) we may be better equipped to understand what the cyborg and cyborgization means to science and society. The cyborg, in a trend toward post-humanism, can be characterized by the ambiguities of both terms. Post- can mean after, and the cybernetic organism can lead humanity into a period of peace and serenity that is configured as post-conflict, post-pathologic or post-poverty. On the other hand, the cyborg may lead to a thresh-

old where the post- collapses into the Latin prefix meaning apart, asunder, away, utterly, or having a privative, negative, or reversing function. In this scenario, the cyborg will un-create, it will deny any form of shared human value(s). The process of the cyborg has begun, and its trajectory and destination rests in our hands.

References

ACHTERHUIS H., *American Philosophy of Technology: The Empirical Turn*. Bloomington, Indiana, Indiana University Press, 2001.

BADMINGTON N., *Posthumanism* (Readers in Cultural Criticism) London, Palgrave Macmillan, 2000.

BAINBRIDGE W. S., BURGER, CANTON, et al. «Expanding Human Cognition and Communication.», in *Converging Technologies for Improving Human Performance. Nanotechnology, Biotechnology, Information Technology and Cognitive Science*, edited by M.C. ROCO and W.S. BAINBRIDGE, 97-178, Arlington, Virginia, National Science Foundation and Springer.

BHUSHAN B., *Springer Handbook of Nanotechnology*. New York:, Springer, 2007.

BOND P. J., CANTON J., DASTOOR M., et. al., «General Statements and Visionary Project. A. Motivation and Outlook.» In: *Converging Technologies for Improving Human Performance. Nanotechnology, Biotechnology, Information Technology and Cognitive Science*, edited by M.C. ROCO and W.S. BAINBRIDGE, 29-96, Arlington Virginia, National Science Foundation and Springer, 2002.

BOSTROM N., *The Transhumanist FAQ*, 2003. At: http://www.transhumanism.org/resources/FAQv21.pdf.

BOSTROM N., «Transhumanist values.» In *Review of Contemporary Philosophy* 4, no.1-2: 87-101, 2005.

CARONIA A., *Il Cyborg: saggio sull'uomo artificiale*. Edizioni Theoria, Roma, 1985.

CLYNE M. E. – S. KLINE., "Cyborgs in Space." In *The Cyborg Handbook*, edited by Chris H. Gray, Jennifer H. Figueroa-Sarriera, and Steven Mentor, 3-42, Routledge, New York, 1995.

GRAY C. H., «An Interview with Manfred Clynes.», in: *The Cyborg Handbook*, edited by C.H. GRAY, J.H. FIGUEROA-SARRIERA, and S. MENTOR, 43-53, Routledge, New York, 1995.

GRAY C. H., J. H. FIGUEROA-SARRIERA, and S. MENTOR, *The Cyborg Handbook*, Routledge, New York, 1995.

HAYLES N. K., *How We Became Posthuman: Virtual Bodies in Cybernetics, Literature, and Informatics*, University Of Chicago Press, Chicago, 1999.

HOLLAND J. H., *Emergence: From Chaos To Order*, Basic Books, New York, 1999.

HOOK C., «Transhumanism and Posthumanism.», in: *Encyclopedia of bioethics*. 3rd ed, edited by Stephen G. Post, 2517-2520, Macmillan, New York:, 2004.

JONES R.A.L., *Soft Machines: Nanotechnology and Life*. Oxford University Press, Oxford, 2008.

MITCHELL M., *Complexity: A Guided Tour*. Oxford University Press, Oxford, 2009.

MUNNIK R., «Donna Haraway: Cyborgs for Earthly Survival?», in: *American Philosophy of Technology: The Empirical Turn*, edited by Hans Achterhuis, 95-118. Bloomington, Indiana University Press, Indiana, 2016.

NOBLE D.F., «Social Choice in Machine Design: The Case of Automatically Controlled Machine Tools, and a Challenge for Labor.» *Politics & Society* 8, no.3-4: 313-347, 1978.

OFFENHÄUSSER A. – R. RINALDI, *Nanobioelectronics – for Electronics, Biology, and Medicine* (Nanostructure Science and Technology), Springer, New York, 2009.

ROCO M. C. – W. S. BAINBRIDGE, *Converging Technologies for Improving Human Performance. Nanotechnology, Biotechnology, Information Technology and Cognitive Science*, National Science Foundation and Springer, Arlington Virginia, 2002.

TURKLE S., «Sociable Technologies: Enhancing Human Performance When the Computer is not a Tool but a Companion.», in: *Converging Technologies for Improving Human Performance. Nanotechnology, Biotechnology, Information Technology and Cognitive Science*, edited by Mihail C. Roco and William S. Bainbridge, 150-158, National Science Foundation and Springer, Arlington Virginia, 2002.

WATERS B., *From human to Posthuman: Christian Theology and Technology in a Postmodern World,* Ashgate Publishing, Surrey, United Kingdom, 2006.

Second part

Workshops

Workshop n. 1

SPIRITUALITY
Neuroscience and the search for "the Person" Case Study: Criminal Law and Human Structure

Antonella Piccinin, Rafael Pou LC

with
Simonetta Baratta, Chiara Marcantonio, Ulderico Moscetta, Anna Pascale, Chiara Santoro, Serena Rieti, Gianmaria Alessandro Ruscitti

Under the supervision
of Prof. Franco Imoda SJ
Pontifical Gregorian University

After a general introduction to 'Neurosciences' and how they have affected other fields, such as philosophy, anthropology, and also law, the aim of this work is to show that there is always a choice for a specific anthropology that is preliminary to any investigation. By focusing on the example of criminal legal system and juridical cases, it is possible to access the pros and cons of different approaches of new questions that arise from the development of neurosciences. Studies have synthesized different reductionist approaches distinguishing between two main views: 1) *dualistic* views and 2) *monistic* views. However, another option is also possible and even preferable: looking at the complexity of the 'human' means also a more 'personalist' approach, open to the a 'holistic' conception of the human being.

1. Introduction: terminology and contents of the study

Today we live in the so-called 'century of the brain' (Yuste – Church 2014). In fact, in the last decades, we have witnessed to a 'neuro-revolution' (Lynch – Laursen 2009)[1] and the proliferation of disciplines with the prefix 'neuro-' (neuro-medicine, neuro-psychology, neuro-linguistics, neuro-ethics, neuro-law, neuro-theology, etc.). All of these disciplines refer ultimately to 'Neuroscience' (or 'Neurosciences') and define what is known as the 'neuro-culture' (Ortega – Vidal 2007), or even a 'neuro-mania' (Legrenzi – Umiltà 2009). This growing interest in this subject "is hardly surprising as neuroscience carries promises of revealing the underpinnings of our individuality, such as emotions, consciousness, the way we make decisions and our socio-psychological interactions" (Frazzetto – Anker 2009: 815). However, the revolution has

[1] Authors refer to a new 'scientific paradigm'. (Bianchi – Gulotta – Sartori 2009: 3).

also caused an ethical and anthropological crisis (Molinaro 2015). The fundamental anthropological question '*Who I am as a person?*' reemerges in light of new progresses in neuro-scientific studies, and will be the fundamental and final theme of this chapter.
The history of *neuroscience* is quite recent:[2] its rapid development started in the second half of the 20th century.[3] Certainly, the milestone for its beginning is even earlier: the discovery of neuron by the Italian scientist, Camillo Golgi (1843-1926), in 1873, that favored the research of the Spanish anatomist, Santiago Ramón y Cajal (1852-1934), and the formulation of the 'neuron theory' (or 'neuron doctrine') (Seybold 2007: 6).[4] The advances of ever more-sophisticated technologies and technological tools have certainly played an important role. However, we do not intend to focus here on technological aspects of this subject, but rather we would like to focus on 'extra-scientific' consequences.
The first question that we need to answer is: *what is* neuroscience? Even though it would not be possible to answer this question briefly, this paragraph will offer the main characteristics of this subject. Firstly, neuroscience is conceived as a *vast* and *hybrid discipline* (Oliverio 2004: 3; Sanguineti 2010: 17), since, it involves many different fields:[5] not only those strictly connected with medicine, chemistry, physics, biology and pharmacology, but also humanistic disciplines, such as psychology, philosophy, law and religious studies (Fabbro 2014). Thus, neuroscience refers not only to one specific area of scientific knowledge or to one specific

[2] However, the name seems to be invented in the 1960s by a neurophysiologist, Francis Otto Schmitt (1903-1995), referring to his inter-university, international, and inter-disciplinary group of research, the *Neuroscience Research Program*, whose main intent was to reach a full understanding of the brain and its functioning (Martucci 2015: 9). See also: Schmitt 1990.

[3] The reason is that only in 1950s there were the first visualizations of synapses thanks to the invention of new techniques, such as the electron microscope and the electroencephalography (Oliverio 2004: 32). Today, there are many different technologies for neuro-, or brain-imaging (Farisco 2012: 21) that are able to see inside the active brain: they are all based on the measurement of the "regional cerebral blood flow", being able to define what area of the brain is active in doing a specific activity. See also: Grandi 2016: 5-10.

[4] Both the researchers won the Nobel Prize award in 1906 (López-Muñoz – Boya – Alamo 2006).

[5] There are at least twenty-four subjects connected to neuroscience.

research field, but to different approaches, with different finalities and methodologies, that investigate the human nervous system (Farisco 2012: 17).

Neurosciences study man and his intellectual reality (and capacities), the functioning of the brain and the nervous system in particular,[6] as well as the link between mind and body; in order to understand man and his actions. Neuroscience – and the connected technologies – have advanced in relation to the understanding of the neurobiological bases of knowledge, feelings and behaviors (Farisco 2012: 9): all of them are "understood as being determined by physical and chemical events in the brain" (Seybold 2007: 18). Thus, the discipline has opened new paths for a more in-depth thinking on issues like the mind, the self, the memory, the emotions/feelings, the personal will/choice, the conscience, and the soul, in order to explain how these elements interact with the organ-brain, or whether they are only a part of it (De Caro – Marraffa 2016).[7] The two main positions – that will be discussed further below – are presented as specific approaches to the subject matter, and the choice for one or the other necessarily is based upon – or delineate – a particular anthropology.

A second matter refers to the possibilities that neuroscience enables: for example, from any form of human (or rather, cognitive) *enhancement* (Savulescu – Bostrom 2009),[8] to the transplant of the brain in

[6] "The nervous system is divided into two basic components: the central and the peripheral nervous systems. The central nervous system (CNS) consists of the brain and spinal cord; the peripheral nervous system (PNS) consists of all neural structures outside the brain and the spinal cord. (…) Everything we perceive, feel, do and think is made possible through processing of electrical messages generated by and carried along these cells of the brain, spinal cord and peripheral nervous systems [the *neurons*]" (Seybold 2007: 7).

[7] The second option is the one favored by "reductionists". See, for example, Changeux 1985.

[8] "Cognitive enhancement may be defined as *the amplification or extension of core capacities of the mind through improvement or augmentation of internal or external information processing systems*. As cognitive neuroscience has advanced, the list of prospective internal, biological enhancements has steadily expanded" (Bostrom – Sandberg 2009: 311; emphasis added). These ideas are opposed by authors, like Sandel, who hold that their finality – especially when it is not addressed to the treatment of mental illnesses (Farah et al. 2004) – is to satisfy a "promethean individualism" (Sandel 2007); moreover these possibilities would result in a new unjust, social disparity between rich and poor people (Lavazza 2010).

machines. The ideas of creating super-men or even to permit immortality do not seem to be only *science fiction* any more (Benanti 2012). Another important issue is: *what influence* does neuroscience have on the other fields, as well as on society and ethical-legal concerns? Neuroscience has supported the rise of relevant philosophical and anthropological, socio-political and economical, ethical and theological, as well as legal, questions. In this chapter, we choose to make a bottom-up path: moving from an analysis of theoretical and practical implications of neuroscience in the law (and criminal law, in particular), we will then show what the most relevant and inevitable philosophical-anthropological consequences are. With the intention of appreciating the undoubtable utility brought by neuro-scientific and neuro-technological researches, we would like to propose an account of risks and opportunities in order to pursue a halfway position: a *personalist view* of man in relation to neuro-technology and neuro-biology.

2. Neuroscience and the law

Deepening the relationship between neuroscience and the law is very intriguing. Even though they might appear to be two very different subjects, they are more increasingly interlinked from different perspectives. Specifically, criminal law (both substantially and procedurally) and criminology are very affected by the application of neuroscience (Flick 2015: 164).
In particular, Sartori – Sanmicheli (2009) have offered a systemization by defining at least three levels of interaction (Martucci 2015: 21; also, Bianchi – Gulotta – Sartori 2009: 17): 1) *Criminal Neurosciences*, that study the neuro-psychic conditions of criminals; 2) *Normative Neurosciences* and of Moral Cognition, that analyses the neuronal basis of the sentiment of justice and morality (Greene et al. 2001; Greene – Haidt 2002); 3) *Forensic Neurosciences*, that are interested in the application of the neuro-scientific evidence in the criminal law trials.
In order to make it clearer, in the following paragraph, we would like to offer a brief description of some legal examples, dealing especially with nn. 1 and 2.

2.1. Cases

Firstly, the possibility of mapping the neuronal activity inside the brain has enabled the recognition of a strict link between some neurological deficiencies and aggressiveness, or other behavioral and mental disorders. A similar approach is the one whose intent is to explain aggressiveness and inclination to violence and criminality from a genetic point of view (Fabbri – Cimino – Serretti 2013). It has been noted that, "in some people, violent behavior has been associated with the presence of dysfunction or injury in the frontal and temporal lobes, whose results are widely documented and that sometimes can lead to manifestations of emotional-behavioral disorder, determined by a reduced control of impulses, but also from a sort of deficit in the ability of learning social and moral norms" (Martucci 2015: 22; our translation). A famous example is the *Case of Phineas Gage* (Lavazza – Sanmicheli 2012: XV-XVI), who after an injury at the frontal part of his head suffered a severe alteration of his behavior and temperament: from being a gentle and calm man, he became rude and aggressive, as much as his physician, John Harlow, affirmed: "*Gage was no longer Gage*" (Damasio et al. 1994: 1102). Such observations were useful in determining the specific area of the brain was linked to a particular activity or function, and how they actually work.[9] However, the possibility of using non-invasive technical tools to study (and mapping) the brain functioning has marked a considerable improvement in this field.

Moreover, medical-scientific development has also permitted scientists to study the interaction of genes on neuronal status: as an example, for a long time (especially in the 60s), it has been debated whether genetic anomalies like the *XYY genome* could cause

[9] Another important study was conducted by the neurophysiologist and psychologist, Benjamin Libet (1916-2007), who analyzed the neural bases of agency. He discovered that the brain initiates conscious voluntary movements as well as the will to move *before* we are consciously aware of the will to move (Libet et al. 1983; Libet 1985). Through his experiment, Libet "suggests that 'the brain "decides" to initiate or, at least, to prepare to initiate the act before there is any reportable subjective awareness that such a decision has taken place' [Libet 1985: 536]. So brains *decide*, or at least 'decide', and *initiate voluntary action*" (Bennett – Hacker 2007: 17; emphasis in original). His experiments supported the 'free will skepticism' (Bayne 2011).

a violent and aggressive behavior or not (Joseph 2001). In fact, many studies – by the British geneticist Patricia A. Jacobs (1934 -), in particular – have been published (Jacobs et al. 1965). It was discovered that a moderate number of criminals with mental disorders detained in British maximum security hospitals presented this anomaly of the sex chromosomes: nine out of three-hundred-fifteen prisoners had an XYY karyotype (according to Jacobs studies), six of which had a much taller and larger stature than the average (Price – Whatmore 1967). The relationship between criminal behaviors and XYY karyotype, today is scientifically unfounded (Tiepolo 2007: 452). It has been discussed in legal and criminal law at the medical levels because the question was whether an individual with this karyotype could be considered irresponsible for his actions, also in the case of a crime.[10] The problem, therefore, includes the ethical sphere more than the purely scientific.[11] From the scientific point of view, it is only possible to say that, under certain environmental factors, an individual with a large build and with a mild mental retardation is more inclined to violence, but this does not imply that all people with such a (or similar) genetic dysfunction are necessarily criminals, as some genetic-cerebral, deterministic positions seems to affirm.[12]

Secondly, from a forensic point of view, in the twentieth century, neurosciences have integrated the traditional methodologies of study of mental processes with various techniques of neuro-scientific analysis, used also for forensic examinations. There are two main instruments that we would like to mention here: the *lie-detection* and the *memory-detection*.[13]

---

[10] Similarly, see the *Case Bayout* (Corte d'Assise di Trieste, n. 5/2009) in which – for the first time in Europe – the Court applied a reduction of punishment for a person accused of murder, referring to the expertise according to which, due to the presence of a genetic variation in the DNA, he would have been more inclined to violent behavior (Lavazza 2010: 98-99; also: Picozza et al. 2011: 55-56).

[11] All these theories recall the Italian physician and psychiatric, Lombroso (1835-1909), and his studies, published in the treatise *Criminal Man* in 1876. See: Gibson 2002.

[12] For further considerations, see: De Caro – Lavazza – Sartori 2013.

[13] These tools, however, cannot be considered as neuroscientific techniques *stricto sensu* (Grandi 2016: 12), but their development has certainly followed the advances in neurosciences' research. An important scientist who supported the validity of these techniques was the psychologist of the University of Minnesota, David T. Lykken (1928-2006),

The first one is an upgraded version of the *polygraph*, which indicates a variation in the blood pressure of the person that is interrogated. Specifically, "using SPM-based [Statistical Parametric Mapping] analyses of multi-subject average group data, several recent fMRI studies demonstrated differences in brain activation between truthful and non-truthful responses in various experimental paradigms" (Davatzikos et al. 2005).[14]
The other tool that has been recently used is the 'memory-detection', whose utility is to identify "memory-traces" and implicit-inconscious information. In order to verify the correctness of autobiographical memories, the *(autobiographical) Implicit Association Test* (IAT) and the *Timed Antagonistic Response Alethiometer* (TARA) have been proposed.[15] Both of these instruments are developed on an algorithmic basis. The evaluation of the memory-content is based on the trace of the time of reaction in computerized tests: fake memories are supposed to produce lower reactions than the true ones.
Today, the use of such methodologies for obtaining criminal evidences poses the question of the validity (as well as the ethical value) of these tools, as they are not yet considered fully reliable, but rather they are still "controversial methodologies" (Dominioni 2005).[16]

---

that in 1959, "reported an experimental study of a new type of lie detector test that he termed the 'Guilty Knowledge Test' (GKT). (…) Lykken's inaugural study demonstrated the technique to be highly effective in distinguishing between knowledgeable ('guilty') and naïve ('innocent') participant tested regarding their involvement in alternative mock crime scenarios: forty-eight out of forty-eight participants (100 percent) regarding a scenario in which they *had not* participated were correctly identified as innocent; forty-four out of fifty (88 percent) tested regarding a scenario in which they *had* participated were correctly identified as guilty. Thus, the mean accuracy of the test across guilty and innocent suspects was 94 percent." (Patrick 2011: 3; 4).

[14] "Researchers have indeed experienced that [the *Functional Magnetic Resonance Imaging* (fMRI)] would enable with and average over 90% to say when a person tells a lie, based on the fact that two brain areas (the dorso-lateral frontal cortex and the anterior cingulated cortex) are more activated when the subject does not tell the truth" (Picozza et al. 2011: 259; our translation).

[15] They were used in the case of "Cremona trial", in which a professional was accused of sexual harassment of a young trainer who worked in his office (Trib. Cremona, 19 July 2011, n. 42588).

[16] In fact, if there is a quite extensive use of these techniques in the US, there is still a resistance of their acceptance in the Italian juridical system, that however has had a pioneering role in Europe. For an extensive analysis of legal cases in Italy, see: Grandi 2016.

2.2 Human free choice: What model for the law?

If from one side, these scientific advances help to better define where is the limit between 'guiltless-unconscious illness' and 'guilty-conscious behavior', on the other side, if brought to their extreme, they would also risk to nullify the principle of subjectivity of the law and personal conviction, that is based on the 'capacity of discernment'.[17] This would also undermine the concept of human 'free choice', and recognize only a biological basis of volition (Martucci 2015: 15-16).

Taking the Italian criminal law system as an example, it is articulated in two main elements: punishment both as burden and as re-education and the requirement of *personal responsibility* of the person judged guilty (art. 13, Italian Constitution). But one can be considered 'responsible' for his own choices and actions only as long as he is 'capable', if he is 'free'[18] in his act of willing (*capacity for rationality*).[19] The person, in order to be punished for his actions (against the law), has to be considered able of representing reality, of understanding the consequences of his actions, and of controlling and determining himself.[20] Thus, "the offense is considered as *a conscious violation* of the criminal rule" (Picozza et al. 2011: 46; our translation, emphasis added). The question that should be answered is: Could A, in his mental neuro-psychological conditions at the time of the deed, not will to perform the action B? (Bianchi – Gulotta – Sartori 2009: 6). Differently from morality, according to the law, the cognitive volition is not punishable in itself, but only in the case in which it

[17] Human personal liberty takes two main forms: freedom of action (without any physical – internal or external – limitation) and freedom of will (as the possibility of choosing between different options) (Martucci 2015: 15). See also: De Caro – Lavazza – Sartori 2010; De Caro – Lavazza – Sartori 203.

[18] The 'choice' must be understood as the ability to withstand the psychological impulses through a conscious discernment. See: Fiandaca – Musco 2014.

[19] Gazzaniga (2005) suggests that more than a 'free will', we should talk about a 'free won't'. In this sense, the person is not totally free of an auto-determination, but he is always free of controlling and stopping certain impulses formed in the brain.

[20] This is the 'classic theory of imputability', that understands the person as a rational and free being, and it was opposed by the "positivistic theory", that holds a biological determinism: each volitional phenomenon is the result of a series of necessary, prior conditions. See: Picozza et al. 2011; Antolisei 2016.

is expressed in an external, criminal behavior (Mantovani 2015). Moreover, any choice must be made by a person that is able of self-determination, otherwise there is no guilt, as it happens in the case of mental illnesses (Garofoli 2013).[21]

Thus, a scientific drift that, through the brain mapping, claims to foresee human behavior and, as a result, the criminal attitude of a person, subjecting him to coercive measures solely on the basis of his own (presumed) biological predisposition to crime, would result in a repression of the person based on variable considerations, in relation to a specific society or historical times. Furthermore, if human beings were to be regarded as governed by a mind as a mere sum of chemical-biological elements (just a 'bunch of neurons and electric impulses') whose interactions are predictable by means of mathematical models and through the use of technological tools, criminal law, as it is today, would no longer have any reason for existing.[22] No punishment would have any meaning, as nobody would be 'responsible' and no crime would exist, but only illnesses and pathologies.

The human being would be so deprived of his free will and reduced to a 'software'. He could be subjected to upgrades (namely, enhancement as well as eugenics) in order to eliminate all the behaviors considered as unacceptable in a given historical moment. In this sense, the negation of human's 'free will' seem not to be based on neuroscience *per se*, as a purely scientific conclusion, but rather a more philosophical argumentation, based on specific philosophical premises that are still under discussion.

Hence, rejecting the theory according to which there would be no real control of man on his own actions, the contribution offered by technology and science is fundamental in allowing to properly identify the alterations that prevent or restrict a person's free will, and in avoiding an infliction of unreasonable punishments where purely medical and neuro-psychiatric interventions are instead required.

---

[21] See: Cass. S.U. 8 Marzo 2005, n. 9163 (also known as the 'Raso Case').

[22] Some authors, like Greene-Cohen (2004) and Sapolsky (2004), suggest a move from a retributive system to a consequentialist one, which is based on determinism and not on 'free will' of the person (Lavazza 2010: 100). Other studies reject the idea of the existence of the 'free will', or rather 'conscientious will', see: Wegner 2002.

3. Neurosciences: Different approaches

In philosophy, Ricoeur recognizes two types of simplification/reduction (Ricoeur – Changex 2002): 1) the dualism body-mind (but also brain-mind/soul); and 2) the materialistic monism.
The first approach is strictly linked with the Cartesian idea of a relation between mind and body, but the two remain different and separate elements: mind, that is immaterial, and body, that is material. Today, this discussion takes the different form of mind-brain relationship (Goetz 2005; Popper – Eccles 1981). However, the same problem still remains unresolved: How can a non-physical element (mind/soul) affect/stimulate a physical element (body/brain)?
The impossibility to answer this question from a dualist perspective has encouraged other possible alternatives, one of which is the 'monist approach', according to which "there is only one kind of thing, which is physical" (Seybold 2007: 90). This view presents many different forms such as: *eliminative monism* that states that there is no mind, only the brain; or also, the mind/soul – and consequently also the self – *is* the brain); or *soft-kind of monism*, that "proposes that while the brain is clearly necessary for mental states to occur, a particular mental state is not identical to a particular brain state" (Seybold 2007: 90).[23] The specific philosophical, monist-material-reductionist position we will discuss here is *scientism* (or rather, *neuro-scientism*).
In all these proposals underlies a scientist and reductionist model of the interpretation of reality, a scientist model that is presupposed and given as obvious. However, this is a stance that we cannot take for granted. This kind of implicit assumptions is unfortunately a largely spread phenomenon nowadays because oftentimes scientists deduce, more or less consciously, philosophical consequences from their scientific discoveries, presenting as scien-

[23] We might also mention the *epiphenomenalism*, as a soft dualism, or as an epistemological dualism that coexists with an ontological materialistic monism. Epiphenomenalism recognizes the existence of the consciousness as an irreducible phenomenon, but denies its causal power, and therefore logically reduces it to a sort of illusion, like the rainbow. For further considerations, see: www.plato.stanford.edu.

tifically confirmed what are nothing but debatable interpretations of experimental data.

Hence, scientism presupposes that natural sciences are the only trustworthy source of truth against the 'soft' truths of philosophy or 'common sense', which are but inferior forms of rationality, and called to disappear as scientific knowledge grows and dissipates the darkness of ignorance in which we still live (Churchland 1989). Neuro-scientific reductionism, therefore, is not the claim that neurosciences can tell us something significant about morality, feelings, thinking and self-consciousness, or the identity of man; rather it is the claim that neurosciences tell us all there is to say about these phenomena. According to this approach, these and other phenomena are (or will be) fully explained by neuronal positions and chemical substances dancing in our brain, and those that are not 'reducible' should be simply denied.[24]

However, this general principal begins with a foundational error: as science cuts out a part of reality, it selects a piece of available information, elaborates a method according to certain goals, and discards the rest not because it is unreal, but because it is irrelevant to its goals. The problem is that "methods become mentalities" (Schumacher 1989: 126), and a partial knowledge finalized to the domination of nature (*technique*) is exalted over time as the only-valid/true knowledge. It is an all-encompassing and totalizing knowledge that crushes and cuts out other areas of reality, bringing them into its 'bed of Procrustes' (think, for example, of the reduction of morality and love to oxytocin) (Churchland 2012), the reduction of thought to electric shocks, the reduction of meaning to neuronal positions). It is also an error because it implies ignoring that science itself is based on presuppositions (whether of a logical-mathematical type, or of an empirical type) that are given to and not produced by science. The crisis of the Vienna Circle and the subsequent evolution of the Philosophy of Science should be able to teach us that lesson, at least. Ignoring one's own pre-

[24] This kind of implicit assumption is unfortunately a largely spread phenomenon nowadays because oftentimes scientists deduce, more or less consciously, philosophical consequences from their scientific discoveries, presenting as scientifically confirmed what are nothing but debatable interpretations of experimental data.

suppositions can only lead to an uncritical assumption of them, and that is the mistake to which scientist reductionism often succumbs. Lastly, it is an error because as an epistemological affirmation (only science gives us knowledge), it is self-contradictory. This, in fact, is not a scientific statement.

To briefly summarize this perspective, we would say that *materialism is the necessary reversal of scientism*: in fact, if scientism is an epistemological thesis (only Science gives us knowledge), materialism is an ontological thesis (there is only matter).

Naturalism (which had in W. Quine one of its most famous proponents)[25] is the epistemological project that seeks to explain the totality of reality based on the natural sciences. Reductionism, in turn, is the thesis according to which reality is composed of simple 'bricks' (called atoms, quarks, molecules or whatever) with stable properties. Complex realities (such as living beings, or thinking beings) are nothing but the result of a 'sum of the parts'. Knowing the laws that govern the simple 'bricks' of reality, it is possible explain all the phenomena of the world, however complex they may be, without having to resort to alternative forms of knowledge. It is the mindset that assumes that, ultimately, psychology is reducible to biology, biology to chemistry, and chemistry to physics (Muci 2010).

Thus, we can say that the materialist interpretation of the mental, in its many different forms, is characterized by the attempt to affirm the thesis that, in the words of Sandro Nannini, just as we can do without God to explain life, we do not need the soul to explain the mind (Nannini 2002). Faced with the dilemma presented by the thesis 'Physics is *causally* complete' and 'mental phenomena are *causally* effective', the materialist draws the conclusion that 'mental phenomena *are* physical'. *Everything that exists is physical*, and therefore mental states are, ultimately, equivalent or reducible to physical-neuronal states.

---

[25] Quine proposed that reality should be identified and described through the natural sciences (understood as well as the social and exact sciences) and not by means of Philosophy, from which it rejected its supremacy and role as the foundation of Science. This philosophical naturalism led him to the project of a *naturalized epistemology*, which would have the purpose of answering all the substantial questions of knowledge and meaning, using only the methods and instruments proper to the natural sciences. See: Quine 1995.

Obviously, this reduction presents not a few difficulties: so materialism in its most radical version, *eliminativism*, simply chooses to deny the existence of the mental. This would belong to a naïve vision of the world, and simply does not exist. Mental states are like the movement of the sun around the Earth, an illusion born from pre-scientific spontaneous knowledge (Churchland 1989 and 1999).

Naturally, this diagnosis of the materialistic position is somewhat simplistic, and ignores that – within the scientist-materialist spectrum – a wide range of proposals and nuances coexist with eliminativism (such as: materialisms of identity, epiphenomenisms, emergentisms, etc.). However, it can offer an adequate idea of the problem, inasmuch as all these versions, in the end, share the scientific and materialistic presupposition, which leads them to affirm that only the material *really* exists, that only the scientific knowledge is *real* knowledge, and that only physical-material causality is *real* causality.[26]

In any case, the explanation of the mental through the tools of the natural sciences appears to the materialist as the only way of defending that the mental is not something "miraculous", inconsistent with the scientific worldview. It is this is the strength and the reason of the apparent success of this model: the crisis of the dualist model, which we will mention later.

3.1. Critics of the reductionist approach

There are numerous fields of discussion that call into question the materialistic model of Philosophy of mind (the *qualia* problem, the propositional contents, the intentionality, and so on...). However, it can be said that, in one way or another, all or al-

[26] The context of this analysis forces us to simplify our discussion, for this reason we have decided to focus more specifically on the deterministic materialism, which is often identified with the materialism "tout court". We are aware, however, that a larger work would have to study also another kind of theories, as – for example – the indeterministic materialism, which proposes that, if our world is not deterministic, materialism can be compatible with free will. If this is proven as true, however, we will have to ask whether indeterminism and freedom are the same thing, or how they can be related. For further considerations see: Weatherford 1991.

most all pivot around the same axis, which is the irreducibility of consciousness. The center of the thesis is that subjectivity is not reducible to a scientific explanation of the material infrastructures that make it possible. That is to say, the subjective experience of being a self, the experience of living 'in first person', the existence of an inner 'movie' of which I am both spectator and actor, the existence of the contents and experiences that are associated with this experience (thoughts, feelings, decisions, etc... the 'materials of the film', so to speak) are not fully explained by the existence of the brain. Explaining the components of the celluloid does not give us a full explanation of what a movie is or why it exists; and explaining the film does not explain the existence of someone who sees and directs it. The appearance in the world of a subjective point of view, of a *self* is a qualitative leap that is not explained by a greater quantity or variety of chemical elements.

The scientific-method's problem of the existence of 'selves', the existence of 'subjective points of view', was emphasized at the time by Thomas Nagel in his famous article *What is it like to be a bat?* (1974). After him, many other authors have held that the progress of the cognitive sciences in itself does not provide the answer to the problem of the relation between brain activity and the genesis of the phenomenological world of subjective experience *in first person*. Knowing the physiology of a bat does not give us even the faintest idea of "what is like to be a bat", "what effect it produces" to be a certain type of creature. As Nagel points out in replying to objections, saying that water is H_2O, or that lightning is an electric shock, is to offer a scientific explanation that explains the appearance as an effect upon our senses of a deeper "reality", but the comparison does not seem to explain the appearance of mental events from physical neurological states, for in this case the subjective "appearance" is precisely the essential character of the possession of a conscious mind.

The 'thought experiments' are also well-known. They illustrate the impossibility of crossing this border: 'Mary's room' or the 'Mary the neurophysiologists thought experiment' (Jackson 1982), the Chalmers' 'zombies argument' (Chalmers 1996), the 'China brain thought experiment' (Block 1990) and the 'Chinese room argument' (Searle 1980). All of them show that we still do not know how a physical mechanism, by the simple fact of realizing certain functional states, can hold qualitative states, and generate a *self*. And even in the dubious case that neuroscience found a clear correspondence between a physical state and a mental state, materialism would not be the necessary conclusion. As Juan José Sanguineti points out in the conclusion of an article:

> We do not know if each of our thoughts corresponds to a specific and differentiable complex neuronal activation. Probably not, but suppose it were so. Well, even in this case there would be no motives for monistic materialism. This supposed neuronal activation would be, in any case, an expression or revelation of thought, which contains it symbolically and at the same time transcends it, just as it happens in any word or phrase said with meaning. And that expression would be understood only by someone who knew how to read it. We are faced with an elevated body, personalized, which manifests the spirit (...) 'Supervenience' is not understood with a purely chemical or electronic concept of the human body (Sanguineti 2003:12; our translation).

The same problem is presented by a functionalist-computational theory of the mind. Thus, even if theorists who present brain as a 'software' perfectly explained the functioning of the brain, the origin of subjectivity and the existence of a 'point of view', of a 'self', would remain ultimately unknown, or at least not yet fully demonstrated.[27] As Juan José Sanguineti points out:

> Even if a robot did physically the same as a man does, and even if it could computationally solve all problems and guide its behavior (computational vision, computational ear, etc.), it would not really *feel*, and it would not have vital, sentient, and personal operations. It would always be a machine, even if it could solve mathematical and logistic problems, simulate emotions or elaborate some works of art. Searle, in particular, has made a powerful critique of the theory of strong artificial intelligence. Computer machines, for Searle, have a derived, non-intrinsic intentionality. Their meanings arise only in

[27] However, we must take into account the many possibilities that the development of thought and scientific research will bring in future, letting us know, understand and do things that we cannot even imagine now. On this issue, see §3.

> relation to users endowed with intrinsic intentionality: human persons. (Sanguineti 2008; our translation).[28]

Nor do we solve the problem, according to Michele di Francesco, with the explanations of some philosophers of the mind or neuroscientists such as Dennet or Gazzaniga, who try to 'de-compose the ego' into different 'cognitive agencies', thus showing that the subject is an illusion, the result of a process. According to Di Francesco, this contrasts too much with our subjective primary experience. The subject is recognized as continuous, solid and well structured: it finds internal divisions and tensions, but within a unitary picture in which there is an 'I' that decides (*I* am the one who decided to study even though my brain is tired, not my neurons). Even demonstrating the existence of these 'sub-personal agencies', nothing would prevent the affirmation of the causal efficacy of the self and its irreducibility to them. When this unit is lost, we speak of pathological cases, very different from ordinary reality (Di Francesco 1996).

[28] On the complex topic of robotics, a recent resolution of the European Parliament (February 16th 2017) has stated as follows: "[the Commission] stresses that the development of robotic technology should aim at integrating human capabilities rather than replacing them; considers essential, in the development of robotics and artificial intelligence, to ensure that men at all times maintain control of intelligent machines; considers that special attention should be paid to the potentiality of emotional attachment between men and robots, in particular vulnerable groups (children, the elderly and the disabled), and stresses the questions posed by the serious emotional and physical impact that such an attachment could have on people" (our translation). While on the one hand the benefits of robotic are numerous, on the other hand, the risks and the ethical implications that they can derive are equally remarkable, with particular reference to its use in the medical field for the 'treatment' and the 'improvement' of humans. This opens the path to the problem of civil responsibility of intelligent robots as it is tied to their intrinsic and necessary concept of 'autonomy'. This character would prevent them from being considered as a mere instrument in the hands of other actors and makes it necessary to reflect on the adequacy of the current rules on civil liability. According to the resolution, it is appropriate to develop new principles and new rules to clarify the legal liability of the various actors for actions or omissions attributable to robots. These considerations are explicitly taken into account in the resolution where extensive spaces have been devoted to the definition of general and particular guidelines. For further analysis on this topic, see § 3.

3.2. Which anthropology for a 'neuro-culture'?

At this point, it is clear that it should be avoided any view that, dealing with 'the human', does not take an account of the complexity and totality of 'the person', characterized by a constant interrelation between brain-body-environment (Farisco 2012). In this sense, both reductionist views – the monist ones (namely, those that consider man only from his biological, or neurological, aspects: so that human choices, feelings, preferences, and behaviors should be explained only from an analysis of cerebral structures) and dualistic ones (namely, those that make a strict distinction between brain-body and mind-soul, as if they were two different elements in correlation but not necessarily in unity) – are inadequate and problematic. But a third option is still possible: a holistic conception of man (Possenti 2006). This conception understands man as *person*, conceived as a unified totality of body, mind and spirit/soul, whose corporeity represents the privileged place of its realization: the person not only *holds* a body, but he *is* his body. It is, in fact, a constitutive element of the person. It is not an objective element, which could be used or exploited in view of an end, but a constitutive dimension of the person, inherent to human nature, lived and experienced by the subject as a manifestation of his identity (Lacroix 1997: 165-207; Seifert 1986: 326-347; Wojtyla 2002: 905). For this reason, any scientific discovery should always be welcome also from an ethical-philosophical point of view, as it would contribute to a better understanding of what 'the human' is. But assumptions – as those put forward by some neuroscientists – that do not hold a 'wide-understanding' of the human as 'a person', then, might be seen even as a threat to the centrality and *dignity* of man, which has always been a distinguishing trait of Western culture. As already affirmed above, it might seem that they lead us to a *radical reduction of man*, considered as a simple product of nature and, as such, not really free, and thus, stripped of his most profound dignity.[29]

[29] Similarly, ethics would be brought within the confines of relativism and utilitarianism, with the exclusion of every moral principle that is valid and binding in itself, thus causing a sharp and profound break not only with Christianity but also with all the religious and moral traditions of humanity (Sanna 2002 and 2003).

As Sanguineti stated, in fact, "neuroscience is teaching us a lot about *what we are* and it especially shows us the significance of our physical (brain) dimension in human behavior up to a previously unknown degree, also in the emotional and cognitive dimensions. 'A lot' does not mean 'all': neuroscience is partial. It does not explain everything" (Sanguineti 2010: 18; our translation). The complexity of the person is not only because he is formed by many different elements, but also because he is presented as a multi-level being, which needs to be analyzed from different perspectives: an *inter-* and *multi-disciplinary* approach (Sanguineti 2010). Thus, dealing with 'the person', we need a thought that is capable of unifying, organizing, linking and interpreting all his (biological, cultural, social, ethical and religious) components and being in dialogue with scientific discoveries: a philosophical thinking on the person should be joined and nourished by scientific achievements (Signore 2010: 116).

To make clear this point, we can start from the distinction between the so-called *top-down* and *bottom-up causes*. The latter assume the kind of causality upon which, as already mentioned,[30] the reductionist mentality is based: reality is composed of simple 'bricks' and complex realities are nothing but the result of a 'sum of the parts'. However, not only philosophical reflection, but the evolution of science itself seems to suggest that it is impossible to explain the whole of reality only on the basis of *bottom-up causes*, and that it is necessary, instead, to recognize the presence of wholes that organize their parts, which seems to suggest the existence of *top-down causes*. Using Strumia's words:

> Recently, that is, since physicists and mathematicians have systematically taken the study of nonlinear systems – a study begun with Poincaré and then abandoned for decades after his death – and since was born the science of complexity, which gradually involved all the sciences with its problems, the reductionist process has experienced a setback, and the relationship between physics and biology has changed radically. (…) The fact that non-linear differential equation in the sum of two or more solutions generally are not a solution constitutes the mathematical basis of the reductionism' crisis, as it avoids

[30] See §3.

> to de-compose a solution that describes a complex structure, or the 'whole' in simpler solutions that describe the isolated 'parts'. This non-reductionist, first and primary character of nonlinear systems of physics can be found in all sciences (Strumia 2002; our translation).

As Robert Muci concludes:

> In conclusion, the reductionist framework is increasingly excluded and so its 'upwards' criterion of causality for the explanation of all levels of life is making space to a 'downwards' criterion of causality (*top-down causality*) that can be scientifically fully assumed. And this criterion legitimate non-reductionist, but holistic frameworks (Muci 2010; our translation).

Actually, this approach turns out to be heuristically fecund in Philosophy of the mind, since it allows us to understand the human being as a system with diverse ontological levels that interact with each other: a system in which a drug can alleviate the headache or stimulate the attention (*bottom-up causality*), but in which if one wants to learn English or makes the decision to overcome a drugs-addiction, a decision of the will will also be necessary (*top-down cause*), which cannot be replaced by a bottom-up cause.

This distinction between *top-down* and *bottom-up* causality has also shed light on psychiatric practice. We can think, for example, of the proposal of Fuchs, according to which mental illnesses should not be treated in a reductionist or unilateral way, but taking into account both the biological and the psychic dimensions. The cure must be sought with the freedom of the patient, through a 'circular' causality – which coincides with what we have been explaining thus far – known as the *Cognitive Reconstructuring* (or also 'Cognitive Reframing').[31] This type of therapy is especially necessary because some mental illnesses, such as addictions (pornography, gambling, etc.), have been provoked by top-down causality (are derived from actions freely performed by the person), although later they might have become biologically rooted, generating true slavery with physical effects comparable to substance

[31] This is part of the more general branch of *Cognitive Behavioral Therapy* (CBT) and it is also recognized as a part of *Rational Emotive Behavioral Therapy*. For further considerations, see: Clark 2013.

addictions (Fuchs 2011). Again, in such a case, a 'materialist' will be forced to say that a psychopath or a gambling-addicted person is such because he has a certain cerebral configuration, while the 'inclusivist' – without denying the possible influence of material factors bottom-up – affirms that a gambling addict is such fundamentally because his decisions have configured that cerebral state. This intuition, lastly, may be a suggestive invitation to rediscover the (unjustly despised and forgotten) *Aristotelian causes*, and to understand how, together with the material and efficient causes (in which, we might say, the natural sciences have legitimately focused), a philosophical consideration of formal and final causes can shed new light to understand phenomena such as life and thought, in a symbiosis of scientific and philosophical perspectives. As Aristotle said:

> One must avoid both the error of "dialectical" (Plato) - for whom, being the soul separable from the body, anger must be fully identified with the "desire for revenge" - and the error of the pre-Socratic naturalist, for whom anger coincides with the boiling blood around the heart. The dialectic, in explaining the movements of the soul and human action, looks to the intentions, that is, the 'real causes' of action, to the form, the "concept" (*logos*), but forgets that the form cannot exist except in a given matter. A naturalist makes the opposite mistake (quoted in: Nannini 2002: 7, 8).

This conceptual tool can also be a valuable aid in order to understand human phenomena such as free will, if we understand the causality of the will, not as a 'thing' that moves other 'things', but as a *formal cause* (as a *top-down* cause) avoiding the reductions already mentioned in this chapter. In this line of thought, we can find proposals of more open-minded anthropologies and philosophies of the mind, as, for instance, the *uni-dual Thomistic anthropology* of Juan José Sanguineti, or James Madden. Their anthropologies could be understood as of the 'incarnate spirit', which equally propose the transcendence of intelligence over its infrastructures ('spirit') and the inseparable psychophysical unity of the human being.

Sanguineti, in particular, points out – as a key intuitions that Thomism can offer to contemporary reflection – the *transcendence of intelligence, will and freedom* with respect to the organic character of the human body, and the understanding of the human soul as the 'form' of the body, as its substantial act in unity with the physical dimension (correlative to a philosophy of sensation and perception in which physical and psychic elements are intimately united) (Sanguineti 2007: 31).

The first aspect helps us to understand why man, unlike animals, is able, among other characteristics, to understand relations and contents in a totally universal way (not only perceiving the usefulness of a thing, but also understanding what usefulness is). The human being has a disinterested contemplative capacity; he has free will (enjoys a potential independence from any finite reality), and also a universal symbolic and technological power. These characteristics of human intelligence, although some of them may appear obvious, cannot be definitively demonstrated or confirmed by any empirical research. It is the philosophical view on reality that reveals these characteristics of our intelligence.

The second aspect (substantial unity of soul and body) allows us to understand how the affirmation of this transcendence does not deny the role of the brain: according to Sanguineti, the intellectual power cannot operate unless preceded by a sufficiently broad and mature superior sensible activity (Sanguineti 2007: 82). This activity must be continuous, in a cerebral function of support of sensitivity constantly open to intellectual activity. The use and development of thought require the creation of language, which is the work of sensitivity in contact with reason. It is thought that creates language, and not vice versa, but creates it out of necessity (Sanguineti 2007: 112). Thought cannot advance without symbols (as we can learn from the example of the so-called wolf-children).

All of this also explains why brain dysfunction and injury can compromise the human capacity to think, act freely, etc. since it shows us that neural conditions are the necessary, but insufficient, material cause: they are not the whole explanation, just as the material causality of the paper and the fingers are not enough to explain how a book is made.

In short, we want to propose an anthropology that refuses to speak of a 'soul that moves the body', but instead speaks of different psychosomatic levels that interact with each other (Sanguineti 2007: 147), and human acts with different dimensions (neurological, psychic, objective, behavioral, metaphysical ones) that we can synthesize by saying that every conscious human act is a single psychosomatic act, with a formal (psychic) dimension and a material (neurological) dimension (Sanguineti 2007: 56).

4. Conclusions: the brain as a necessary but not sufficient condition.

The statement "the brain is a necessary, but not sufficient condition" (Merleau-Ponty 1967) can easily synthesized what we have offered in this chapter. Similarly, the *Memorandum* 'Reflective *Neurowissenschaft*' (2014)[32] affirms:

> Ultimately, the reduction of the human person and all his intellectual and cultural performances to his brain as a 'new image of man' **is totally insufficient**. In this unilateral scheme, it is not possible to conceive *the human being as subject and person* in his multidimensionality.[33]

What should be opposed is the ideological tendency to a 'naturalization' (or 'biologization')[34] of the human being, which – in the case of neuroscience – becomes equating man and his brain: thus, man *is* his brain, and only it. We move from the *personhood* to the *brainhood* (Vidal 2009: 5): man is just a 'cerebral subject'.[35]

---

[32] This document was presented by a group of leading German neuroscientists after ten years from the publication of the *Manifesto* on neuroscience, written in 2004 with the intent of offering some premises. According to the *Memorandum*, none of the premises have been satisfied (Müller 2015: 658).

[33] *Memorandum* 2014 (our translation; emphasis in original; bold added). See: www.psychologie-heute.de.

[34] It refers to the idea that it would be possible to give a complete scientific-natural explanation of human thinking, feeling and acting. See, for example: Sperry 1981; Crick 1994; Gazzaniga 2005.

[35] "The ideology of the cerebral subject involves a fundamental paradox. The brain appears as the material organ par excellence, as the only part of our physical bodies that is really worth exercising in order to improve ourselves" (Ortega – Vidal 2007: 258). But, as Lavazza points out, there no full scientific evidence to support such a position (Lavazza 2010: 95).

However, if neurosciences help to understand the functioning of the human brain, explaining that "the brain is involved in all of our behavior and experience" (Seybold 2007: 77), recognizing the fundamental role of feelings and emotions in making decisions (Haidt 2001), and overcoming the Cartesian dualism between body and mind, they do not necessarily demonstrate (or support) any *determinism* in human action and choice (Flick 2015). In this sense, a personalist approach – as we have suggested here – is more inclined to recognize the deep complexity of human being, conceiving him as an 'incarnate spirit' (Lucas Lucas 2005). By preferring this view, we refuse to choose between either man in his biological-material existence, or man in his spiritual reality: rather, the two components (and not only these two!) *are* the man in his totality. If we believe that the man cannot be 'reduced' to his brain *only*, this does not mean that we cannot understand that the man is characterized *also* by his brain (as neurosciences explain). Therefore, in order to understand the multifaceted nature of the human being, it is necessary to hold a wide horizon, in which science can play its fundamental role, but always in relation with other disciplines, for the formulation of a complete account of man as an holistic and dynamic being. This wide view of man ensures also the recognition of his intrinsic *dignity*. Scientific and technological developments are useful means to understand that man is always more than what we know and experience through science, and this becomes even clearer if we think that also "the brain – and the way it gives rise to conscious thought – remains one of the great mysteries in all of science" (Yuste – Church 2014).

References

ANTOLISEI F., *Manuale di Diritto Penale*, Giuffrè, 2016.

BAYNE T., «Libet and the Case for Free Will Scepticism», in: Swinburne, Richard (ed.). Free Will and Modern Science. Oup/British Academy, 2011.

BENANTI P., «Cyborg and Cyborgization», in: Giordano, James (ed.). *Neurotechnology: Premises, Potential, and Problems*, CRC Press, 2012.

BENNETT M. – HACKER P., «Philosophical Foundations of Neuroscience», in: Bennett, Maxwell et al. (eds.). *Neuroscience and Philosophy*. Columbia University Press, 2007.

BIANCHI A. – GULOTTA G. – SARTORI G., *Manuale di Neuroscienze Forensi*, Giuffrè, 2009.

BLOCK N., Excerpt from «Troubles with Functionalism» in: Lycan W.C. (ed.), *Mind and Cognition: A Reader*, Blackwell, 1990.

BOSTROM N. – SANDBERG A., «Cognitive Enhancement: Methods, Ethics, Regulatory Challenges». *Science and Engineering Ethics*. Vol. 15 (3): 311-341, 2009.

CHALMERS D. J., *The Conscious Mind: In Search of a Fundamental Theory*, Oxford University Press, 1996.

CHANGEUX J. 1985. *Neuronal Man*. Princeton University Press.

CHURCHLAND P., *A Neurocomputational Perspective: The Nature of Mind and the Structure of Science*, MIT Press, 1989.

—— *Matter and consciousness: A Contemporary Introduction to the Philosophy of Mind*, MIT Press, 1999.

CHURCHLAND P., *Neurobiologia della morale*, Cortina, 2012.

CLARK D.A., «Cognitive Reconstructuring», in: Hofmann, Stefan G. (ed.) *The Wiley Handbook of Cognitive Behavioral Therapy*, Vol. I. Wiley-Blackwell, 2013.

CRICK F., *The Astonishing Hypothesis. The Scientific Search for the Soul*, Macmillan, 1994.

DAMASIO H., et al., «The return of Phineas Gage: clues about the brain from the skull of a famous patient», in: *Science* Vol. 264 (5162): 1102-1105, 1994.

DAVATZIKOS et al., «Classifying spatial patterns of brain activity with machine learning methods: Application to lie detection», in: *NeuroImage*. Vol. 28 (3): 663-668, 2005.

DE CARO M. – LAVAZZA A. – SARTORI G., *Siamo Davvero Liberi? Le Neuroscienze e il Mistero del Libero Arbitrio*, Codice, 2010.

—— *Quanto Siamo Rsponsabili? Filosofia, Neuroscienze e Società*, Codice, 2013.

DE CARO M. – MARRAFFA M., *Mente e Morale*, LUISS University Press, 2016.

DI FRANCESCO M., *Introduzione alla filosofia della mente*, Carocci, 1996.

DOMINIONI O., *La Prova Penale Scientifica*, Giuffrè, 2005.

FABBRI C. – CIMINO L. – SERRETTI A., «Genetica dei Comportamenti Impulsivo-Aggressivi: Possibile Applicazione in Psichiatria Forense?», in: *Rassegna Italiana di Criminologia*, Vol. 1: 34-51, 2013.

FABBRO F., *Neuroscienze e Spiritualità*, Astrolabio, 2014.

FARAH M. J. et al., «Neurocognitive enhancement: what can we do and what should we do?», in: *Nature Reviews Neuroscience.*,Vol. 5 (5): 421-425, 2004.

FARISCO M., *Filosofia delle Neuroscienze*, Edizioni Messaggero, 2012.

FIANDACA G. – MUSCO E., *Diritto Penale Parte Generale*, Zanichelli, 2014.

FLICK G.M., «Neuroscienze e Diritto», in: Sangalli, Samuele (ed.). *Futuro e Responsabilità*, LUISS University Press, 2015.

FRAZZETTO G. – ANKER S., «Neuroculture», in: *Nature Reviews Neuroscience*, 10 (11): 815-821, 2009.

FUCHS T., «Are Mental Illnesses Diseases of the Brain?», in: CHOUDHURY, SUPARNA – SLABY, J. (eds.), *Critical Neuroscience*, Wiley-Blackwell, 2011.

GAROFOLI R., *Manuale di Diritto Penale*, NelDiritto Editore, 2013.

GAZZANIGA M.S., *The Ethical Brain*, Dana Press, 2005.

GIBSON M., *Born to Crime: Cesare Lombroso and the Origins of Biological Criminology*, Praeger, 2002.

GOETZ S., «Substance Dualism», in: GREEN, JOEL B. – PALRMER, STUART L. (eds.), *In Search of the Soul: Four Views of the Mind-Body Problem*, InterVarsity Press, 2005.

GRANDI C., *Neuroscienze e Responsabilità Penale*, Giappichelli, 2016.

GREENE J. et al., «An fMRI Investigation of Emotional Engagement in Moral Judgment», in: *Science*, Vol. 293: 2105-2108, 2001.

GREENE J. – HAIDT J., «How (and Where) Does Human Judgment Work?» *Trends in Cognitive Science*, Vol. 6 (12): 517-523, 2002.

GREENE J. – COHEN J. «For the law, neuroscience changes nothing and everything» *Philosophical Transactions of the Royal Society: Biological Science*, Vol. 359 (1451): 1775-1785, 2004.

HAIDT J. «The Emotional Dog and Its Rational Tail: A Social Intuitionist Approach to Moral Judgment», in: *Psychological Review*, Vol. 108: 814-834, 2001.

JACKSON F. «Epiphenomenal *Qualia*», in: *Philosophical Quarterly*, Vol. 32: 127–136, 1982.

JACOBS P. A. et all., «Aggressive Behaviour, Mental Sub-normality and the XYY Male», *Nature*, Vol. 208: 1351-1352, 1965.

JOSEPH J., «Is Crime in the Genes?», in: *The Journal of Mind and Behavior*, Vol. 22 (2): 179-218, 2001.

LACROIX X., *Il corpo di carne. La dimensione etica, estetica e spirituale dell'amore*, EDB, 1997.

LAVAZZA A., «Dalle Neuroscienze alla Neuroetica», in: Renna, Luigi (ed.), *Neuroscienze e Persona: Interrogativi e Percorsi Etici*, EDB, 2010.

—— *Il Delitto nel Cervello*, Codice.

LEGRENZI P. – UMILTÀ C., *Neuro-Mania. Il Cervello non Spiega Chi Siamo*, Il Mulino, 2009.

LIBET B., *et al.*, «Time of unconscious intention to act in relation to onset of cerebral activity (Readiness - Potential)», in: *Brain*. Vol.106: 623-642, 1983.

LIBET B., «Unconscious cerebral initiative and the role of conscious will in voluntary action», *Behavioral and Brain Sciences*, Vol. 8: 529-566, 1985.

LÓPEZ-MUÑOZ F. – BOYA J. – ALAMO C., «Neuron Theory, the Cornerstone of Neuroscience, on the centenary of the Nobel Prize award to Santiago Ramón y Cajal», in: *Brain Research Bulletin*, Vol. 70 (4): 391-405, 2006.

LUCAS LUCAS R., *Man Incarnate Spirit*, Cicle Press CT, 2005.

LYNCH Z. – LAURSEN B., *The Neuro Revolution*. St. Martin's Press, 2009.

MANTOVANI F., *Diritto Penale*, CEDAM, 2009.

MARTUCCI P., *Neuroscience e Processo Penale*, Key, 2015.

MERLEAU-PONTY M., *The Structure of Behaviour*, Beacon Press, 1967.

MOLINARO J., «L'umano Fallibile e le Neuroscienze», in: *Concilium*, Vol. 4: 716-731, 2015.

MUCI R. «Riduzionismo e neuroscienze», in: RENNA L. (ed.), *Neuroscienze e persona: interrogativi e percorsi etici*, EDB, 2010.

MÜLLER K., «Vaniloqui a Proposito di una Nuova Dottrina Salvifica?», in: *Concilium*, Vol. 4: 657-670, 2014.

NAGEL T., «What Is It Like to Be a Bat?», in: *The Philosophical Review*, Vol. 83 (4): 435-450, 1974.

NANNINI S., *L'anima e il corpo: un'introduzione storica alla filosofia della mente*, Laterza, 2002.

OLIVERIO A., *Prima Lezione di Neuroscienze*, Laterza, 2004.

ORTEGA F. – VIDAL F., «Mapping the Cerebral Subject in Contemporary Culture», in: *Electronic Journal of Communication Information & Innovation in Health*, Vol. 1 (2): 255-259, 2007.

PATRICK C. J., «Science on the Rise: Birth and Development of the Concealed Information Test», in: VERSCHUERE, BRUNO – BEN-SHAKHAR, GERSHON – MEIJER, EWOUT (eds.), *Memory Detection*, Cambridge University Press, 2011.

PICOZZA E. et al (eds.), *Neurodiritto. Una Introduzione*, Giappichelli, 2011.

POPPER K.R. – ECCLES J.C., *The Self and Its Brain*, Springer International, 1981.

POSSENTI V, *Il Principio-Persona*, Armando, 2006.

PRICE W.H. – WHATMORE P.B., «Behaviour Disorders and Pattern of Crime among XYY Males Identified at a Maximum Security Hospital», in: *British Medical Journal*, Vol. 1: 533-536, 1967.

QUINE W., «Naturalism; Or, Living Within One's Means», in: *Dialectica*, Vol. 49 (2-4): 251 - 263, 1985.

RICOEUR P. – CHANGEUX J., *What Make Us Think?*, Princeton University Press, 2002.

SANDEL M.J., *The Case Against Perfection*, Harvard University Press, 2007.

SANGUINETI J.J., «La mente inmaterial en Santo Tomás de Aquino», in: International Congress *L'Umanesimo cristiano nel III milenio: la prospettiva di Tommaso d'Aquino*, 2003.

—— *Filosofia della mente: una prospettiva ontologica e antropologica*, EDUSC, 2007.

—— «Filosofía de la mente», in: FERNÁNDEZ F. – MERCADO J.A. (ed.), *Philosophica: Enciclopedia filosófica on line*, 2008. http://www.philosophica.info/archivo/2008/voces/mente/mente.html, (last access: 30/12/2016).

—— «I Compiti delle Neuroscienze e l'Antropologia Filosofica. Un Discorso sul Metodo», in: RENNA L. (ed.), *Neuroscienze e Persona: Interrogativi e Percorsi Etici*, EDB, 2010.

SANNA I., «Il futuro dell'uomo e i nodi della questione antropologica», in: *Nuova Responsabilità*: 88-10, 2002.

—— «L'identità umana nell'epoca della globalizzazione», in: *Rivista teologica di Lugano*, Vol. 8 (2): 369-388, 2003.

SAPOLSKY R.M. «The frontal cortex and the criminal justice system», in: *Philosophical Transactions of the Royal Society: Biological Science*, Vol. 359 (1451): 1787-1796, 2004.

SAVULESCU J. – BOSTROM N. *Human Enhancement*. Oxford University Press, 2009.

SEARLE J. «Minds, Brains and Programs», in: *Behavioral and Brain Sciences*, Vol. 3: 417–57, 1980.

SEIFERT J., *Essere e persona. Verso una fondazione fenomenologica di una metafisica classica e Personalistica*, Vita e Pensiero, 1989.

SCHMITT F.O., *The Never-Ceasing Search*, American Philosophical Society, Vol. 188, 1990.

SCHUMACHER E., *A guide for the perplexed*, Abacus, 1989.

SEYBOLD K.S., *Explorations in Neuroscience, Psychology and Religion*, Ashgate, 2007.

SIGNORE M. «Neuroscienza e Libertà della Persona», in: RENNA L. (ed.). *Neuroscienze e Persona: Interrogativi e Percorsi Etici*, EDB, 2010.

SPERRY R., «Changing Priorities», in: *Annual Review of Neuroscience*, Vol. 4: 1-15, 1981.

STRUMIA A., «Materia», in: TANZELLA G. – STRUMIA A. (eds.), *Dizionario interdisciplinare di Scienza e Fede*, Urbaniana University Press, 2002.

TIEPOLO L., *Le Basi Biologiche dell'Ereditarietà*, Goliardica Pavese, 2007.

VIDAL F. «Brainhood, Anthropological Figure of Modernity», in: *History of Human Sciences*, Vol. 22 (1): 5-36, 2009.

WEATHERFORD R., *The Implications of Determinism*, Routledge, 1991.

WEGNER D.M., *The Illusion of Conscious Will*, MIT Press, 2002.

WOJTYLA K., «Persona e Atto», in: REALE G. et al. (eds.), *Metafisica della persona. Tutte le opere filosofiche e saggi integrativi*, Bompiani, 2002.

YUSTE R. – CHURCH G.M., «The New Century of the Brain», in: *Scientific American*, Vol. 310 (3): 38-45, 2014.

Workshop n. 2

POLITICS
Technology and Democratic Procedures Case Study: How technology shapes democracy in the social networks and "post-truth" era

Nino Marzullo, Nicola Pigna,
James Cleary LC

with
Federico Poggianti, Marco Iervese,
Alessio Piazza, Simone Di Giovambattista,
Jacopo Solaini, Giorgio Simone

Under the supervision
of Prof. Giovanni Moro
Università di Roma Tre
Pontifical Gregorian University

1. Thesis

"The internet today catalyzes the fear that man experiences whenever his history is intruded upon by those technological revolutions capable of changing his relationship with his environment" (Nicholas 2011: 317). With this phrase, we start these reflections on the influence that the digital revolution exert, not only by delivering content but, above all, by its capacity to shape users' choices.

This latter point is especially evident in the phases of the decision-making process of three electoral processes, namely the Italian and British referendums and the Colombian plebiscite. The widespread use of cable television, the vast possibilities that the internet offers, its evolving uses and the overall success of social networks are some of the factors that have more forcefully influenced the change in the relationship between the public sphere, the media and citizens. For these reasons, one can speak of 'disintegration in political communication' that has brought about a transformation in the decision-making processes of the citizens (Ceccobelli 2014: 657).

Just last century, the relationship between citizens and the public institutions, with the necessary mediation of the political parties, had a certain established nature. This relationship was based on processes of grouping and organizing, favored and controlled by the bureaucratic apparatus of these same unions and political parties. The diminishing presence of such a system has highlighted the difficulties which the traditional political parties manage or relate with the diverse technologies which the web has made available to voters. The electorate, with a growing digital know-how, feel more involved with personal causes, specific movements or spontaneous gatherings that spring up and develop from the internet.

The changes to the traditional approach political communication have resulted in innovations in campaigning and political communication, especially on the eve of highly contested electoral decisions. Papini argues that the advent of new technologies allows us to speak of post-communication: that is, when communicating via these technologies becomes the major sounding board as to the message, and indeed amplifies it more than in the past: it is a communication that has an increasing influence on electoral choices (Papini 2011).
This influence should not be overlooked. Monitoring the general sentiments expressed on social media can lead to changes in decisions on the communication or programming of electoral campaigns, and with the possibility of making predictions based on the opinions expressed by these sentiments on the digital platforms. Della Porta's research has shown a substantial difference between the myth of direct democracy – a term more and more popular among promoters of social media - and its most widespread application that is taking on a different character, that of deliberative and participatory democracy. This deliberative and participatory democracy sustains that the preferences of the participants can change in the course of the interaction.
This concept goes well beyond the idea of a bourgeois public sphere. It favors a broader understanding, an alternative public arena, in which resolution mechanisms become important (Della Porta 2011). And it is precisely in such alternative spheres that the social movements move. These movements, born on the internet, promote the idea of a political participation that is much more inclusive than that of the traditional political parties. Sorice maintains that new technologies of communication form a set of tools of great importance for the development of a representative space of the new social movements. This does not, however, imply that the Internet is responsible for the growth of such movements. It would be rather simplistic to consider that social media, by itself, can foster a "revolution" (Sorice 2016).[1] Yet, as Morlino notes,

[1] "The «new» communication technologies (in particular those deriving from the web 2.0) constitute an important set of tools and, at the same time, a symbolic representative space for the development and affirmation of new social movements. We certainly cannot

digital platforms are an indispensable instrument to ensure the expression of disadvantaged groups' voice, thus significantly improving the quality of democracy (Morlino 2014).
In addition, it should be pointed out that social media has altered users' preferences, so much so that authoritative studies have shown that the brain has a greater ability to adapt itself than what was previously assumed: "According to the theories of neuroplasticity, the genes establish many connections between the neurons, constructing the structure of the base of the brain. But experiences determine the long-term efficacy of these connections: in this way, the continual reconfiguration of the mind becomes possible in addition to the elaboration of new behavioral patterns." (Barisione – Blasio – Hibberd – Higgins – Sorice 2013: 394) Because of the large quantity of information, ideas and emotions to which our brain is subjected when we have to make a choice, it is increasingly clear how necessary it is to reflect on the time variable as an element capable of determining a greater or lesser awareness of choices. Time dedicated to deciding is directly proportionate to the quality – awareness – of the choices.
For decades, cognitivist psychology has shown that the social reality, be it in it individual aspects or in its collective dimension, is characterized by a grade of complexity that makes it difficult – if not impossible – to know the social reality accurately. Such difficulty has been noted for two reasons, be they "quantitative reasons (the stimuli that we receive are excessive in comparison to the well-known capacity of our mind to them), or for qualitative reasons (many of them are indeed ambiguous, complex and multifaceted)" (Giacomini 2016). Therefore, in addition to the previously mentioned time variable, there is another: we might call it the 'emotional' variable. This is generated by our emotions, our prejudices and our expectations. Often, political decisions are conditioned by these variables and end up not coinciding with reasonable choices, reasoned, formulated and optimized for the

say that the Internet is responsible for the growth of the movements, just as we cannot say that social media alone can «make» a revolution (Gerbaudo 2012). At the same time, however, we cannot but recognize the growing importance of the web (and, more generally, of the media) in the development of new forms of democracy." (Sorice 2016).

"average citizen", who distractedly gets involved in the choice of ready-made solutions.

This chapter's analysis springs from this outlook and from the relevance that the internet has today. This is backed up by a June 24 news article published by media around the world, which is both of great influence and very timely: The referendum on the United Kingdom remaining in the European Union had just ended. The British people had voted just a few hours earlier to "leave". Then a Google Trends tweet revealed an especially important fact about the referendum. This tweet, accompanied by a chart, showed what the search results relating to the theme "EU" were from the moment news broke that Britain was officially leaving the European Union (*La Repubblica*, 24/06/2017). Among the top five search results stood out questions like "What does it mean to leave the EU?" or "What is the EU?", or "How to get an Irish passport?". These searches increased 100% after the announcement of the result. The British, or at least the British who had carried out searches on Google, had recently left their voting booths, yet there were still many doubts about what they had voted for. These questions revealed an obvious fact: we are increasingly dependent on the web and less informed, or at least ill-informed from reliable sources.

It is due to these unreliable and less than truthful sources that a whole host of websites and false online news stories have been able to proliferate. Indeed, the so-called "fake news" is an ever-growing phenomenon, especially in the last year.

The constant increase in fake news is provoking an ever-greater misinformation. Every time news of this kind is spreads, on the web or through traditional media, it is difficult to deny it or at least try to correct the falsehood of what has been said (The Guardian, Viner 04/06/2017). During the campaign of the UK referendum "Brexit", fake news played a decisive role in the final victory of the "Leave" party, which promoted the country's exit from the European Union. The news spread by the leader of the "Leave" campaign regarding the country's economic benefits of leaving the EU, was one of the most shared stories on the web and had a strong appeal among the British electorate. Only at the end of the referendum did the truth about this news come out: the

entire claim was false and was merely a pretext to tip the argument in favor of leaving the European Union (The independent, Stone 11/06/2017).
Starting from this pretext regarding Brexit, and the influence that the internet and the social networks had on the result, we have reflected on the actual valence of referendums as instruments and on events, both important in their respective countries, which here we mention briefly: the Italian referendum on a proposed Constitutional reform and the Colombian plebiscite concerning a peace treaty between the state and the FARC.
Here it would be opportune to mention an important and pressing theme: the evolution of the role of social networks in our society and how this evolution relates to these three previously mentioned referendums. Therefore, we will rely on analysis regarding the multi-faceted uses of the internet that have recently evolved. From this, the question leads to the possibility to analyze the data of how many have a Facebook or Twitter account so as to be able to capture the so-called sentiment of the internet. The general sentiment has shown that the internet is always more and more an accurate mirror of our society, above all as to voter preferences.
Today, because of the vast amount of data to analyze and compare, as well as the presence of countless variations caused by varying political and social systems, it is almost impossible to make definitive or accurate evaluations which establish the degree that new technologies impact electoral choices.
The internet's central role has ended up with it being the instrument of growth for organizations that are defined as "fluid", that is, suited to mobilizing and involving voters which no longer result from complex and formal party structures - like those of the past century - but rather spring from sharing information that "provokes" results that were, perhaps surprisingly, sometimes already well underway.[2] A certain misunderstanding, common to journalistic over-simplification, must be avoided. This simplification treats these forms of political action as if they lacked organization.

[2] One must remember the protest movements such as "Occupy Wall Street" and groups that mobilized sectors of voters offering alternatives to traditional political parties such as Five Star Movement in Italy and *Podemos* in Spain.

On the contrary, the lack of these traditional organizational elements, a fact that becomes increasingly clearer as they take on new forms, allows for a greater speed in communication because these forms do not have to worry about the logistics of management that were historically burdensome and costly (Shirky 2009: 21). Today, most of politics is conducted on the internet and above all on social networks. The following pages will serve to show what occurred in the three referendums that have been the subject of analysis and as well to point out the expanding nature of this phenomenon: the impact of social networks on the private life of individuals, and in this case specifically on their voting intentions, as their political opinions often are expressed on the internet.
We have observed how the internet, and especially social networks such as Facebook and Twitter, have collected a very high amount of information and opinions regarding the referendums. However, the political debate on social networks and information shared on the internet does not yet play the decisive role in the political decision of people. Our paper seeks to emphasize, therefore, only the influence of new technologies on the voting and their ability to reach a very high number of people in the world.

2. Note on methodology

In the following pages, the reflections center on three important referendums – as they are recent – and on the mainstream media's perception of the influence that the digital information has on the decisions of the citizens in these referendums. The methodology followed uses sources from the countries where the referendums were conducted, analyzing the principal positions of different cultural sensitivities and expressions of individuated social interests. Those identified and described in this chapter are of a quantitative analysis which describe the framework for the influence of the social networks in Italy, the United Kingdom and Colombia but also takes into account the major newspapers reporting in these countries and their political leanings.
The methodology of this paper has undergone several modifications: Initially it contemplated studying the mainstream media's opinion

regarding the influence that of the social media in the three referendums. This was done by comparing the first 50 results of the two major newspapers of diverse political leanings of the country where the referendum was held. The results were registered in a database and categorized with both quantitative and qualitative questions. To do this, the method used the search function of the newspaper's websites using the same keywords to ensure valid results.

Given the limited results returned, the methodology moved to comparing the first 10 pages of Google search results using the same keywords established. This allowed for a wider scope to include blogs and television and radio. From there, a qualitative analysis of the articles would categorize the impressions that each article had of social networks' influence in the electoral outcomes. However, many articles were reporting of the same source material such as studies conducted, as reporting styles differ depending on the country and its cultural expectations of mainstream media. Thus, they did not emit opinions on social networks influence but merely reported the studies' conclusions.

Therefore, this paper's methodology developed again, going now directly to these sources, the studies themselves.

One should not fail to point out how much moving beyond the pre-social network dimension of the political discourse has represented an opening up of this discourse to interested parties and the users of the information. This new social dimension is ever more evident due to the possibility for interaction, which is significantly increased because of this new situation, which the internet offers to voters. An aspect to be highlighted of this new dimension is that of the instruments that this new form of communication used. Studies and the raw statistics of social networks show the influence in the election period and in predicting the outcome of the vote.

Each of these three cases represents very different cultural and political contexts. Additionally, the types of referendum were binding in diverse manners, according to the laws of the country in question. Yet these variables do not prevent the possibility to evaluate and analyze the response of the voters to both the referential instances, witnessed by high turnouts (except in the case of Colombia) which indicate an interest that is higher than other

elections and the desire to participate that the internet has fostered and even heightened.
In the analysis, it was found that the social networks play the role which one could call "reinvigorating" for an electorate that perhaps was disenchanted with the political process and unable to make well-informed choices.
An insight that can be gleaned from studying these mainstream newspapers under the light of the three referendums, is that in analyzing every facet of the electoral campaigns, as always done by the serious journalism, they also highlighted an important new phenomenon. For the first time the impact of social networks in shaping voters' decisions especially in the run-up to voting was a studied as an important area of study, especially in regards to the results of the diverse campaigns in social media, with different approaches and political structures. The overall picture that developed allows one to not only understand how the political parties are changing, but also to evaluate the broadcasting capacity of these media, such as Facebook and Twitter, which are able to reach millions of people, with at times "fake news", as is referred to by the word "post-truth" communication.
The first attempt to influence the vote using social networks was by Barack Obama in the presidential elections of 2008. His campaign organized a strategy of creating awareness and they were the first to use alternative methods to the then normal political communication, such as town-hall meetings, TV debates. Obama's campaign took advantage of the new opportunities that the internet offered (Effing 2011: 25-35). The results of this strategy showed how using Facebook allowed them broadcast short and insightful messages to a large number of persons and how this was able to shape public opinion.
Obama, having himself used social networks to reach millions of people and keep them up-to-date about his activities in his two presidential terms, wanted to dwell on the topic of post-truth and denounce its danger. The former President acknowledged the new ecosystem of news as something in which "everything is true and nothing is true" (The Guardian, Woolf 11/06/2017) and called on the main social networks to intervene to stem what increasingly is a serious problem.

Yet, the dilemma of post-truth and the widespread use of social networks places us in a new position: We are living in unique period in regards to both freedom of expression and the large number of people in the world who have access to this news. All of this, however, has a limit: today there is too much information and it is difficult to distinguish what is true and what is not, what can we establish as reliable and what is a fake – a figment of the imagination.
At the same time, in our paper, we have also highlighted how the role of new technologies increases the polarization of the electorate towards one of the two options on the referendum's ballot paper. The voter's motivations can simplified due to a pressure that a growing number of "public spaces" exert on the users by projecting a "majority" opinion and not promoting dialogue. These spaces, while indeed beneficial, sometimes lack an assessment of their merits. This assessment is necessary, however, for the effective functioning of other direct democracy tools. Here we refer to popular initiatives or popular assemblies, where the citizen has a much more meaningful participation, having to propose a more thought-out assessment of the choice they make or the proposed legislative text they chose.
The research here encountered the aforementioned phenomenon in all three case studies as will soon be outlined. In each of the scenarios analyzed, public opinion and the outcome of the vote was influenced by social networks, with fake news playing a significant role. These "sources" were often the most shared on social networks and also the most difficult to denounce as false by those who were accused of these unverified facts. This study thus aims to shed light on this phenomenon, with the aim of creating awareness regarding fake news.

3. The constitutional reform in Italy

The first of the case studies presented in this chapter, is that of the Constitutional Referendum in Italy. Held on December 4, 2016, its objective was "the provisions to overcome equal bicameralism, reducing the number of members of parliament, to contain of expenditures of the government institutions, the suppression of the CNEL (National Council of Economy and Labor) and the revision of Title V of Part II of the Constitution."[3]

During the first months of 2016, it involved all political forces, each with different objectives. But what most characterized this referendum from previous ones was the involvement generated on the Internet, especially on social networks, by the political parties, activists and all those who expressed an opinion about using their own "profile". Before, discussions normally took place in the headquarters of political parties, in the squares and major social venues. They are now a distant memory. More and more people are deserting these places, preferring to find their information by watching television or on the social networks. Indeed, one out of three Italian use social networks as their preferred platform to keep up-to-date on the major news stories (CENSIS 2016: 8).

This new situation has brought about a new manner of communicating. According to the latest Census Report, 36% of Italians opt for Facebook and Twitter as their preferred source of news, in comparison with 27% in 2015. The report stated that "the digital technologies have not only changed our manner of communicating, of relating with others, of informing judgments, but have also produced radical changes in many daily ways of acting." (CENSIS 2016: 15-16). One of these changes, the subject of this chapter's analysis, is that specifically of the impact that technology has on politics and the capacity in these last years to connect with and change the public opinion, adapting itself as the times change.

As they are accustomed to do in these situations, the principal Italian mainstream media, with the opportunity offered by the constitutional referendum, have been able to analyze every single aspect of the electoral campaign. Yet they noted a significant new development. For the first time, they analyzed the impact that social networks have on the voters' intentions, considering specifically the consequences that these create, in the pre-voting stage and, having concluded the voting, the results obtained by the different *social* campaigns delivered in different methods by the political

[3] See: "Disposizioni per il superamento del bicameralismo paritario, la riduzione del numero dei parlamentari, il contenimento dei costi di funzionamento delle istituzioni, la soppressione del CNEL e la revisione del titolo V della parte II della Costituzione".

parties. The overall vision that these analyses will allow one to not only understand how the Italian political parties are changing, but will also be useful to have a greater understanding of the capacity for disseminating messages that medium of communication such as Facebook or Twitter have today: they are capable to reaching millions of people, delivering content that is at times "fake news", which derives from the concept of "post-truth".

Before beginning a deeper analysis of the Italian Referendum, it would serve well to distinguish what are the instruments used by companies to explore the universe of posts and tweets that has characterized the months prior to the referendum. An analysis of the research carried out and the data collected on the social networks will be analyzed regarding the predictions about voting on the Constitutional Referendum.

3.1 Tools used and Pre-voting Predictions

The use of social networks by the Italian political parties is a relatively recent innovation. In Italy this type of approach was only adopted a few years ago and the first to bank on the power of digital propaganda were the representatives of the Five Star Movement (M5S), a party / movement that sprung from and fostered by the internet, during the campaign for the 2013 parliamentary elections (Cataldi 2013: 101-103). Subsequently, all parties have taken steps to adapt, implementing various models for publicizing their political thought on the social networks. The first results of this remodeling were seen in the Constitutional Referendum of December 4 from the use that each party has made Facebook and Twitter throughout the election campaign that preceded the vote.

The arguments that the principal Italian political parties offered to the public have been constantly analyzed, as is the case for many decades, by pollsters, using for example a sample analysis. This time, it was possible to count on one more tool: the analysis of the so-called "big data" and, therefore, of the entire flow of information present on social networks, with particular reference to data which covered the referendum.

Blogmeter, a specialist agency that offers social media monitoring solutions, conducted a specific study on the data that comes from

Italian users of Facebook and Twitter concerning the constitutional referendum. This observed the intention of the voters, the social networks more widely used and the keywords that characterized the online discussions. The overall vision that was obtained served to decipher manner of acting of Italians on social networks, but also how the manner of communicating has changed. According to Blogmeter, "in the period from September 24th to December 1st, 4 million public messages about the referendum, posted by more than 800,000 users, which in turn generated 25 million interactions (+140% more to the month before), on social media and 336 million unique views were registered" (Blogmeter 2016). The social network that was most used by Italians was Twitter which accounted for 54% of messages written concerning the referendum, while only one out of three preferred to post their personal opinion on Facebook.

The most surprising result was that the 1.8 million user created messages, allowed to forecast the voting intentions of the Italians. 60% of users studied had published posts in favor of voting "no", versus the 40 % of users who supported voting "yes" (Pennisi 2016). These last statistics when compared with the final result of the referendum, corroborate the analysis performed by Blogmeter and shows at the same time how more and more technology is able to examine the so-called *sentiment* of public opinion (Liu 2012: 7).[4]

The *sentiment* was researched not only by Blogmeter. In fact, other realities in Italy have worked on big data to offer analysis regarding the theme of the "constitutional referendum". Among these, the society Sixth Sense[5] has contributed with its own study. The difference from other analysis already mentioned comes from the different perspective of the study: "Sixth Sense, by means of a pro-

[4] "Sentiment analysis, also called opinion mining, is the field of study that analyses people's opinions, sentiments, evaluations, appraisals, attitudes, and emotions towards entities such as products, services, organizations, individuals, issues, events, topics, and their attributes. It represents a large problem space. There are also many names and slightly different tasks, e.g., sentiment analysis, opinion mining, opinion extraction, sentiment mining, subjectivity analysis, affect analysis, emotion analysis, review mining, etc. However, they are now all under the umbrella of sentiment analysis or opinion mining" (Liu 2012: 7).

[5] Sixth Sense is a spin-off of the Scuola Superiore Sant'Anna di Pisa.

prietary algorithm that analyses tweets available on the internet, has analyzed the trends of the tweets regarding the constitutional referendum within the 2 weeks prior to the voting" (Sixth Sense 2016). Also, in this case, an enormous amount of data was generated and the results led to a different prediction to that of Blogmeter: in the days just before the vote, the "No" was losing ground as the proponents of "yes" in turn were quickly growing. Both points of view offer interesting valuations and forecasts, which allow one to observe the course of the flow of information relating to the constitutional referendum, which proliferated on the social networks in the months before the vote.

3.2 The results of the Referendum, the power of the web and its influencers: post referendum analysis

The predictions and exit-poll, along with a heated debate involving all political forces for months, ended on December 4, 2016, the day when Italians went to the polls to approve or reject the constitutional reform.

This referendum saw a high turnout, 65.47% of voters, and a clear outcome with the 59.12% of the valid votes against the reform, compared to 40.38% of votes in favor and 0.92% voting with empty ballot papers. It is interesting to analyze how the vote was differentiated by geographic areas of the country. The only regions where the "yes" won were Emilia Romagna, Tuscany and Trentino Alto Adige; "No" instead held significant advantage in all other regions. As for the turnout, the first place went to the Veneto region with 76.66% of its registered voters, while on the contrary in Calabria only 54,43% of those eligible voted.

Finally, as regards the large urban zones, in Rome the "no" won with 59.42% of the vote, confirming the results of the regions. This is also identical in Naples and Campania with 65% voting for "no". On the other hand, the "yes" won in Florence and Bologna with 56.29% and 52.23% of the votes respectively. An isolated case is that of Milan where the "yes" barely won with 51.13%, while the surrounding region in Lombardy voted "no" with 55.50% of the votes (*Ministero dell'Interno* 2016).

These are the official figures provided by the Interior Ministry, derived from the official count. Yet they were produced from data according "alternative" valence. Social networks, as argued above, have partly influenced the vote. A careful and analytical reading of the posts published by the Italians makes it possible to get to the final outcome predictions. The framework provided by the Blogmeter, due to its semantic analysis of big data, estimated almost with exactness the outcome of the referendum vote, arguing that the "no" was the position held by 60% of the Italian active in social networks. This result supports the fact that "The internet has long been a true reflection of social dynamics: it is necessary – as is always the case when trying to understand public opinion – to use accurate methods and possess extensive multi-faceted knowledge of the phenomena. The mix of these skills can generate accurate insight on the opinions held by people, even in the political field". While some predictions have proved to be exact, at the same time analysts of social networks stressed that while it was expected a victory for the "no", they did foresee such a vote margin in favor of the "no".
The company Talkwalter carried out a study on the hashtags used on election day and the following day, coming to more than 342 thousand results – and limiting them to just the 5 most used hashtags. This analysis has indeed crowned Twitter as the most used social network within 48 hours of voting, having delivered as much as 98.9% of the multimedia content published on the internet (Russo 2016). The politician most quoted was Matteo Renzi, who generated over 60 thousand search results, making his name among the most "trendy" hashtags that evoked the rejection of his proposal for reform. This last aspect summarizes clearly what the real focus of this referendum vote was: the Italian referendum was a political vote more than being about a constitutional reform or the proposals that arose from this.
Another constant of this referendum were the awareness campaigns implemented by the political groups, in particular those of the Democratic Party (PD), on the one hand, to support voting "yes", and on the other the M5S, a movement in favor of voting "no".

In the first case, it was observed that the PD communication was sterile and too focused on its leader, the only influencer, obscuring those who did their utmost every day to promote the same political cause. The second case, the M5S, campaigned instead with very argumentative tones, being perceived by the public as a threatening and aggressive. However, in the end, what favored the victory of the "no", and thus the success of the M5S (and other political campaigns that participated in this effort), was the ability of the activists to know to be widely present on the main social networks and to interpret and capture the population groups who were most needy at that time and favor their rise and redemption. As an IPSOS dossier observed after 9000 interviews conducted immediately after the end of the referendum, "the 'no' is a young and popular vote: the highest points are among the unemployed and among the workers.
In the case of Italy, the social networks have given the opportunity to be able to predict the intentions of the public opinion's vote and have offered at the same time a new survey instrument.

4. Brexit in the UK

The impact that social media has had on public opinion and on reaching agreement is as much a debated matter in today's scientific literature as it is increasingly complex (Holt 2013: 21-24). With structures, language, and algorithms in rapid evolution, we are witnessing a clear growth of these "systems of aggregation" (Dron 2014) into ever-growing, ever-more-digitalized groups within our population. The case of the consultative referendum held in the UK, which the press has defined as the first "digital referendum" (Mullen 2016),[6] is a clear example of this phenomenon and provides us with a case study of great potential.
In Great Britain, social media was pivotal in the campaign, reducing more traditional means, such as the papers or television,

[6] Both campaigns spent millions of pounds on Facebook and online advertising. The head of Labour In for Britain, a separate campaign to that deployed by Britain Stronger in Europe, confirmed that the Labour Party alone spent over £1 million on Facebook and online advertising and successfully reached 13 million people in the process.

to a supporting role. Furthermore, the limiting of public funds to £600.000 for both parties was an incentive to look for other means of financing, leading to fundraising campaigns which strengthened the parties' Internet presence and influence. This led to notable investments in social media, which were already-structured platforms with an incredible capacity for communication. According to the press, millions of pounds were spent, and this allowed for a greater outreach to more voters (*Il Foglio*, Peduzzi 27/05/2017).

Faced with this new starring role played by the social media, inquiries have been made to understand the real extent of their influence on the voting public. According to a study done by the Electoral Reform Society, 16% of those interviewed considered the social media their most important source of information for the referendum. This number grows to 33% within the age group of 18-24, who preferred social media to other traditional sources of information, such as family members (27%), the BBC (24%), friends (23%), and newspapers (16%). Thus, the social media seem to have been held as one of the most important tools of influencing the voting public, especially the young.

Furthermore, if we consider the record 72.2% turnout of voters (8% more than in 1975 and 30% more than in 2011) we can reasonably conjecture that the massive presence of the campaigns on the social media was indeed an effective attempt at getting people involved. This has led many analysts to wonder what the winning strategy was and by what means Leave's highly unexpected victory was made possible (Porcaro – Muller 2016).

Many studies have shown that the progress made by those in favor of Leave was largely due to their communications strategy in the months prior to the referendum. Contrary to Remain, whose campaign mainly appealed to economics and other practical reasons for choosing to stay, the pro-Leave group brought up points such as security and immigration, placing emphasis on the more inflammatory aspects of the debate. Moreover, the highly alarmist way in which many approached these matters was made clear by the abundance of tags such as "ISIS" and "terrorism" in social network discussion forums (Siegel 2016). It was the more emotional appeal of the pro-Leave campaign that would grab the public's attention, by creating a large number of users that would eventually reach critical mass and decide the fate of the referendum.

However, this strategy gave rise to many doubts about the truthfulness of many of these information sources, as well as, in a broader sense, the dangers of an unfiltered, instrumentalized management of social media. Indeed, some news, such as Nigel Farage's claim that leaving the EU would save Britain a weekly £350 million, were shown to be false. Even so, this did not stop their use as propaganda, creating a notable impact on public opinion and increasing aversion toward Remain. Sentiment analyses (Turney 2002: 417-424; Pang, Lee et al. 2002: 79-86; Pang – Lee 2005: 115-124) have shown an evident prevalence of Leave supporters in the time prior to the vote, beginning in July 2015 with the immigration related crisis. This prevalence goes hand in hand with the spirit of alarmism that was created about the problem. Subsequent news kept the focus more frequently on Leave's positions than on Remain's (Llewellyn 2016). This use of public unrest, riding on peoples' reactions of fear and unease, often by means of non-verified news sources, has been a strategy both dubious and effective. Also, it has been shown to be perfectly adapted to both the type of referendum in question and the medium involved.

In the referendum, the complex spectrum of points of view, a natural part of any election, is simplified to a binary option. This not only simplifies the vote but also acts as a catalyst for many deep and complex sentiments. From anti-politics to anti-establishment to politics of opposition, the referendum is in itself a way of expressing dissent. For the pro-Leave campaign, playing to the voters' emotions had positive effects, involving and committing people to their values more than Remain was able to do. And when the news seemed not to be tragic enough, they were often built upon.

Furthermore, the most popular social networks, such as Twitter, Instagram, and especially Facebook, proved by their very structure to be most useful to such a strategy. Being as they are "filter bubbles" (Bozdag 2013: 209-227),[7] which hardly lend themselves to in-depth knowledge, (Dunne 2016) the information that these platforms provide is moderated by friendships and known interests (The Guardian, Bell 03/0602017). Thus the algorithms that generate news access, or newsfeeds (Constine 2016), supply us with news sources we already know and trust. This puts us at the risk of exposure to incorrect information or of being withheld other sources which, though unknown to us, could provide a critical evaluation of what we think to be true. The apparent "democracy" of information provided by social media risks being a supplier of networks that, for lack of cross-referencing and verification, turn out to be self-referential.

These self-sufficient machines of persuasion generate an "information cascade" in which the credibility of the news is based not on verifiability, but on the blind trust we place in our sources of information. And when certain pieces of news are cleverly manipulated, danger obviously follows.

The Brexit is a clear case of how the use of social media as a tool for propaganda, while highly effective, is also highly problematic, often tending toward unethical. In a system where "post-truth" politics, as Moïsi has called them (Dominique 2008), tend to produce emotional short-circuiting rather than movements of freedom and information, the viral nature of social media communications can be a great risk if not managed correctly.

Referencing Moïsi's concept of post-truth, we can analyze the electoral behavior of Brexit, as with any other political-social variant, through the prism which George Ritzer (*Santambrogio* 2014) proposes. He defines a "McDonalized" society where one is ringed by standardizing behaviors, closed in with affective bonds dictated by society. Such a situation is present in Brexit's run-up. It is clearly evident how each citizen is distanced from that freedom, that which is proper to the man, to the emancipation for which so many have fought and given their lives.

[7] A filter bubble is a result of a personalized search in which a website algorithm selectively guesses what information a user would like to see based on information about the user (such as location, past click behavior and search history).

This desire to stand up (Aristotle spoke of *Zoon politikon*) of man has little by little diminished, being quenched, in an almost unnoticed way, by the totalitarian institution (per M. Facoult's term) (Croteau – Hoynes 2015) of the web where we are unable to determine our actions or where we are unwilling to do so to avoid a true and proper mediatic shaming, for fear of exclusion or complete marginalization. These are reflections on society in general, and thus can be seen in the electoral behavior in Great Britain. While it one could contest to what degree it was a factor in the referendum, we cannot deny that we are ourselves the protagonists in our lives and must, by using social media, take back our subjectivity and our presence of mind to improve not just our current situation but also for that of the generations to come.
It would appear that a fundamental rethinking of criteria for the correct use of these media is in order. Far from making them less open, democratic, and participative, it would maintain and confirm them as such, while striving to fulfill their purpose through the sharing of truthful information.

5. The peace process between the government and the FARC in Colombia

The plebiscite in Colombia was held on October 2, 2016. The purpose of the plebiscite was the ratification by Colombians of the peace agreements (*"Acuerdo final para la terminación del conflicto y la construcción de una paz estable y duradera"* or commonly known as the *"Acuerdos de paz de La Habana"*) between the Colombian government (Gobierno Nacional) and the Revolutionary Armed Forces of Colombia (*Fuerzas Armadas Revolucionarias de Colombia-Ejercito del Pueblo* or FARC-EP). These agreements were signed on August 24, 2016.[8] Only 37.43% of enrolled voters participated in the plebiscite. The "no" won by a slight margin, 50.21% to 37.43% (*Registraduría Nacional del Estado Civil*).

[8] The Final Agreement is available at www.urnadecr stal.gov.co

5.1 The Background to the Conflict between the FARC and the Colombian Government

The Republic of Colombia is a South American country, bordered by Venezuela, Pananama, Brazil, Ecuador and Peru. Colonized by the Spaniards in 1499, it gained independence in 1819 and in 1886 became a constitutional republic. Yet the process of state building set up the conditions for the conflict that brought about the plebiscite (Gonzaléz 2004: 12).

Since the 1960's the country has been in conflict with armed guerrilla groups. But this conflict can be traced back to an earlier period of '*La Violencia*' (1948-1958) which followed the assassination of the popular leader Jorge Eliécer Gaitán Ayala (Gozalez – Mendoza 2016: 22). In an effort to stop the partisan violence, the National Front was formed and approved as a national government by Colombians on December 1, 1957 (Ruiz 2001: 108). Yet this new government was exclusive with power shared between the two predominant parties, the Liberals and the Conservatives. Lack of economic reforms – especially an agrarian one – created the environment of dissident political groups. In addition to this, the revolutionary figures of Fidel Castro and Che Guevara were popular. An outcome was that some dissident leftist political groups resorted to forming guerrilla groups. As a response, rightist paramilitary groups were formed (Tate 2001: 165). While many groups were dissolved gradually after the Constitutional reforms of 1991, the largest guerrilla group, FARC-EP continued.

There have been several attempts to broker peace with the FARC-EP as "the present peace negotiations are the fourth attempt in 30 years to have formal talks with the FARC to end the insurgency." (Beittel 2015: 14) The peace talks that lead to the plebiscite were begun formally on August 26, 2012 under President Juan Manuel Santos.[9] There was a major breakthrough in September 2015 (Gehring – Koch 2016: 96) and they were signed on August 24, 2016.

[9] *Alto Comisionado para la Paz, "Mecanismo de refrendación de los acuerdos."* The Agreement is available at www.mesadeconversaciones.com.co

5.2 The Legal Nature of the Plebiscite

Number 6 of the August 26, 2012 peace document with FARC-EP refers to having a referential mechanism. Article 103 of the Colombian Constitutions lists plebiscite as a form of participation (Coward 2016: 26). While the FARC-EP proposed a 'constituent assembly',[10] the plebiscite was the referential mechanism proposed by President Juan Manuel Santos on August 22, 2013 and approved by Congress on September 17, 2013 (*Presidencia de la Repùblica de Colombia*), a necessity of Article 104. (*El Pais* 17/09/2013)[11] However, this mechanism also required the Constitutional Court to rule on it as per article 241 §3 (*El Tiempo* 24/08/2016). The Constitutional Court approved the plebiscite on July 18, 2016, thus making, according to Article 104, the decision of the people legally binding.

The court set the guidelines of the plebiscite, namely:

(1) bring democratic legitimacy to the final accord's implementation effort; (2) lock the accords in so that future leaders will have to implement them "because popular approval has obligatory character," and (3) give the parties strong guarantees that what was agreed will be complied with. (WOLA 2016)[12]

However, the Court was also clear that the plebiscite for this accord was only binding on the President.

5.3 The Influence of the Social Networks on the Plebiscite

The referendum on the peace agreement with the FARC, promoted by the Columbian government, witnessed a surprising victory of those who voted 'No' to the agreement, thus manifesting a clear

[10] This would be able to change the Constitution per Article 241 § 2, www.uniandes.edu.co and www.telesurtv.net.

[11] This was necessary as article 104 of the Colombian Constitution prohibited referendums from being held at the same time as other elections: Santos had initially hoped the plebiscite would be held during the 2014 elections.

[12] "A potential disapproval of the Final Accord only has an effect on the implementation of this specific public policy decision, leaving untouched the other state organs' competencies, among them the President's ability to maintain public order, including through negotiations with illegal armed groups, with the intention of reaching other peace accords."

opposition to those in favor of sealing a definitive peace between the Marxist guerilla group and the State. This has defied all predictions and has, in some way, surprised not only the institutional leaders, but also the international media and, still more, the Colombian press.
A central motive of the shock, was the total unpredictability of the result which no study foresaw as plausible. Particularly relevant to the scope of this paper is the clear conflict of the traditional press's predictions, which seem incapable of verifying how many of those who expressed opinions in favor of a 'Yes' vote in the weeks preceding the referendum, actually changed position in the end.
It is possible to share the view that articles online mainly offer reports on data collected in conversations that take place in social networks, while few actually analyze the context and correspondence between the themes treated and language used regarding the referendum itself. The results, typically gained by means of automatized methods, have left a gap that has denied the possibility of an analysis to produce effective results.
The results gained by a study commissioned by the Mision de Observacion Electoral (M.O.E.) between 17 June and 30 September of 2016 on Twitter, Facebook and Instagram registered more than two million mentions of the referendum by 417,060 users. Of these, 7% reported offensive or intolerant expressions and 10% expressions of hope. From this, it seems that the level of polarization was relatively low, the most extreme expression not arriving to 20% of the whole. Along with these, only 2% of the users mentioned the risk of fraud. Therefore, on the one hand, there is a majority of comments in favor of the agreement, even if not explicitly argued for, but above all, a widespread confidence of the electorate in the democratic function of the referendum. The same study reported that 83% used completely neutrally-toned language. This made it especially difficult for the traditional media to interpret the trends that may have led to the failure of the peace agreement.
Another very interesting phenomenon is that 12% of all the conversations regarding the referendum was conducted by a mere 60 users. This implies a diffusion of highly mediated content. On the basis of these data, it is evident that the campaign in favor of the

agreement attempts to use a large number of images and charts that, even though at first sight giving the idea of a mass impetus, were later shown to be a weak element, especially because of the greater dissipation of Yes message, which did not succeed to focus their efforts to pass on to the electorate a unified message. (*El Colombiano*, Restrepo 3/10/2016)

In light of these observations, there is a strong implication of the implementation of the social as a means of communication, even if it is clear that this is not capable of being a trustworthy mirror of reality. Particularly, relying still on the study conducted by M.O.E. on Twitter, Facebook, YouTube, Instagram and blogs from 6 to 29 September of 2016, that is, in the final phase of the referendum campaign, the comments of those in favor were surpassed by more than 40% by those against.

The fundamental fact is generational: youth between 18 and 30 years old promoted the peace agreement, especially the age-group of 18-24 years old. These contributed to 54% of the comments regarding the referendum, 80% of which were in favor of the peace agreement. However, the position of those registered as users over 45 years old was quite different: 75% of these, in fact, defended the No vote. The age group of between 30 and 45 years old offered a more balanced position, dividing equally the comments in favor and against the peace agreement. (*El Tiempo* 30/09/2016).

The theme of the referendum, inevitably, has aided a polarization around these two positions, leaving little space for eventual arguments and conversations, thus limiting reasons for support or critiques against. The inevitable hostility that resulted is fundamentally due to the use of exaggerated or dishonest messages (*A la Luz Pública* 2016). The consequent polarization has driven users to use social network to confront and consolidate an already taken position. The aggressive tone and disinformation circulating around social networks, therefore, drove those who hold either opinion to close themselves to dialogue. Thus, the determinant algorithm of social networks, which selectively and automatically imposes what information will be seen by each user according to his expressed preferences, limits the interactions between both sides of the cause, as in the case of Columbia.

Another aspect that emerged during the campaign was the uncontrollable and at times well-employed use of misinformation as a weapon, strengthened by the fact that the media user rarely seeks to verify news sources, with the consequence that numerous accounts incessantly diffused messages to feed doubts about the other side. The traditional media has often pointed out and stigmatized this phenomenon. (CNN *Español*, Solano V., 03/10/2016)
It remains a question yet to resolve as to the reason for the lack of correspondence between the data collected from social media conversations – which pointed in favor of the 'yes' – and the outcome of the referendum. It is difficult to classify this phenomenon. Yet some hypotheses are possible. The dominance of the favorable message towards the peace agreement could be because of strong government intervention – the greatest official promoter of the peace deal – that with a massive intervention in the social networks could have shaped the discourse.
Such action could have been perceived by the Colombian voters as interference in the outcome, in addition to being proof of the weakness of the arguments and the Government's uncertainty of the outcome, and may therefore inevitably diminished its credibility before the electorate. The language used was not much different than that of traditional media: this created even more a feeling of a disconnect, given that the social media are the vehicle par excellence where the message becomes customized. Ultimately, too general messages, combined with a greater propensity of the promotors of 'yes' to stigmatize the opposing opinions with extremely aggressive language, contributed to the victory of those who rejected the agreement, who in the ballot secretly expressed what they dared not publicly declare on social networks. (*La Semana*, Rincón 10/04/2016)
In the case of Colombia, the social networks have been shown to play an important role in the debate, albeit littered with verbal violence and a diffusion of fake news. This calls for a serious reflection on how such an instrument can affect the democratic life of the country and, above all, how real is perception that once can derive from social media a correlation between common mood regarding an issue and opinions published on the internet.

Conclusions and the future

The referendums in this study, while taking into account the undoubted political, cultural and economic differences between individual countries clearly have one thing in common.
The influence of social media was important to determine the outcomes of individual referendums considered for this study. A common factor that emerged from the three experiences can be found in the lack of awareness and an inability, by the main political parties, to use social networks both to convey institutional messages and to repudiate fake news. The internet, therefore, in the three cases analyzed, benefited those political forces that counted on the advantage of being able to leverage the delegitimization of leadership in general and catalyze a protest message that forced institutions in power to devote a lot of energy to highlight what drives its political choices.
In this regard, a new element of the today scenario, which derives from what has just been said, is fake news. It has developed over time and has assumed a dimension in some ways unpublished in relation to the referendums in question. For it is still difficult to establish from a quantitative point of view the effect that the spread of this news has had on the decision-making processes of voters. However, counteracting the spread of fake news has proved to be a very complex and difficult task.
Today, Facebook and Twitter, to name only the better known, are tools that few can do without. Everything occurs, "lives" and is shared on one's personal page, on one's social profile, becoming immediately no longer just a memory or a personal thought, but content of the public domain, and as such, subject to comments of people who are personally unknown. These instruments, born initially to connect the world, are today a mirror of our lives. Because of this, they are analyzed in-depth to such a degree to be able to analyze many different perspectives, in our case the political dimension of users.
No one can affirm with certainty, at this moment, that the social networks have determined the outcome of the three referendums that have been analyzed. Their power to influence the public opin-

ion, however, is extremely clear. Thus the political parties have found a new asset to be used for their election campaigns. From post-truths, to polls and sentiment, the panorama described in these pages shows an ever more quickly changing world. This does not allow anyone to remain behind, especially those who still strive to make their voices heard, even through electronic devices. The challenge for the future will be that of being aware of the means at our disposal and to acquaint the younger generation with the consequences that their social behavior can have. Technology helps man, but it does not replace man in his private life or in the expression of his thoughts.

In conclusion, we can sum up by saying that the social media has had an initial and visible result in substituting the "places" where politics traditionally was found in: the party groups, the newspapers, the provincial and national secretariats, the unions and so many cultural and social institutions that are in some way linked, directly or indirectly, to political activity. Among the causes of this "dematerialization" of the physical places of encounter is surely the greater immediacy that the internet offers to conveying information. As well the communication is made easier by the lack of those heavy "mediators", that often were real and proper filters, such as party newspapers and radio stations, only to mention the most common. The social media have taken on a key role in fundraising since the internet has become an alternative change to many funding campaigns which are often linked to specific electoral needs. The individual groups' capacity for communication has become, indeed, now a way to persuade and convince the entire electorate, while before the traditional media reached were only its followers.

The innovations that were mentioned in this chapter are certainly a starting point for reflection concerning the unforeseeable outcomes that the influence of the internet will have on electoral choices. It is reasonable to believe that the technological refinement will lead to a greater ability to predict the outcome of electoral contests, a fact that should not, however, affect the secrecy of the vote. This is because often the victories of "silent majorities" give pause to reflect on the falsification of the general electoral preference derived from big data.

By studying the instrument of the referendum, analyzed in three different political and historical contexts, we have been able to assert that we are facing a new moment in our history, one that is just beginning - the social networks era. The internet has allowed the world to be constantly connected but even interconnected. Our opinions are influenced daily and political news is everywhere. However today we have the opportunity to interact with political figures and discuss with them as never before. This allows us to inform ourselves better and have a better awareness of what our political choices are. Only time will tell us if this daily and widespread political debate will bring about good results for our society and for future generations.
The reflection is still ongoing and will surely be for years to come.

References

ALTO COMISIONADO PARA LA PAZ, *Mecanismo de refrendación de los acuerdos,* 2016, in: www.mesadeconversaciones.com.

A LA LUZ PULICA, *Redes Sociales: la clave del plebiscito*, 08/08/2016, in: www.alaluzpublica.com

BARISIONE M., DE BLASIO E., HIBBERD M., HIGGINS M. e SORICE M., *La leadership politica. Media e costruzione del consenso in Comunicazione politica*, 3, 2013.

BEITTEL J., «Peace Talks in Colombia», in: *Congressional Research Service*, March 31, 2015.

BLOGMETER, *Il Referendum costituzionale 2016 sul web*, 2/12/2016, in: www.blogmeter.it.

BOZDAG E., «Bias in algorithmic filtering and personalization» in *Ethics and Information Technology*, 15 (3): 209–227, 2013.

CATALDI M. – EMANUELE V., «Lo tsunami cambia la geografia e strappa 50 province a Pd e Pdl» in DE SIO L., CATALDI M. E DE LUCIA F. (a cura di) *Le Elezioni Politiche,* 2013.

CECCOBELLI D., «Andrew Chadwick. The Hybrid Media System: Politics and Power» in *Comunicazione Politica*, Vol 3, ComPol, 2014.

CENSIS, *50° Rapporto sulla situazione sociale del Paese/2016*, Franco Angeli, Roma, 2016.

CNN *Españo*l, Solano V., «Colombianos, atrapados en la red social de la polarización», 03/10/2016, in: www.cnnespanol.cnn.com.

EL COLOMBIANO, RESTREPO V., ¿Por qué en las redes sociales ganó el Sí?, 3/10/2016, in: www.elcolombiano.com.

CONSTINE J., *Facebook puts friends above publishers in "News Feed Values" and ranking change*, 29/06/2016, in: www.techcrunch.com.

COWARD M., HELLER P., (et al), *Colombia's Constitution of 1991 with Amendments through 2005*, Oxford University 2016, p. 26, in: www.constituteproject.org.

CROTEAU D. – HOYNES W., *Sociologia generale*, 2015.

DELLA PORTA D., *Democrazie*, Il Mulino, Bologna, 2011.

DOMINIQUE M., *La géopolitique de l'émotion: Comment les cultures de peur, d'humiliation et d'espoir façonnent le monde*, Flammarion, Paris, 2008.

DRON J. – ANDERSON T., *Teaching crowds: Learning and social media*, Athabasca University Press, 2014.

DUNNE H., *Did social media call Brexit?*, 31/07/2016, in: www.corpcommsmagazine.co.uk

GEHRING H. – KOCH M., «On the Path from Failed State to OECD Member?», in: *International Reports* (32), Konrad-Adenauer-Stiftung, 95-110, 2016.

GIACOMINI G., *Psicodemocrazia. Quanto l'irrazionalità condiziona il discorso politico*, Mimesis, 2016.

GONZALÉZ F., «The Colombian conflict in historical perspective» in: *Accord* 14, 10-15, 2004.

THE GUARDIAN, BELL E., *The truth about Brexit didn't stand a chance in the online bubble*, in: www.theguardian.com.

HOLT K., SHEHATA A., STRÖMBÄCK J., and LJUNGBERG E., «Age and the effects of news media attention and social media use on political interest and participation: Do social media function as leveller?», in: *European Journal of Communication*, 2013.

LLEWELLYN C., *The results are in and the UK will #Brexit: What did social media tell us about the UK's EU referendum?*, in: www.referendumanalysis.eu

LIU B., *Sentiment analysis and opinion mining. Synthesis lectures on human language technologies*, 2012.

MINISTERO DELL'INTERNO, *Riforma costituzionale 2016: il Referendum*, 2016.

MORLINO L., *Democrazia e mutamenti. Attori, strutture, processi*, LUISS University Press, Roma, 2014.

MULLEN A, *Leave versus Remain: the digital battle*, 2016, in: www.referendumanalysis.eu.

NICHOLAS C., *Internet ci rende stupidi? Come la rete sta cambiando il nostro cervello*, Raffaello Cortina Editore, Milano, 2011.

EL PAIS, *Congreso aprueba proyecto de referendo por la paz propuesto por Santos*, 17/09/2013, in: www.elpais.com.co.

PANG B. – LEE L., «Seeing stars: Exploiting class relationships for sentiment categorization with respect to rating scales», in: *Proceedings of the Association for Computational Linguistics (ACL)*, 2005.

PANG B.; LEE L. – VAITHYANATHAN S., «Thumbs up? Sentiment Classification using Machine Learning Techniques», in: *Proceedings of the Conference on Empirical Methods in Natural Language Processing (EMNLP)*, 2002.

PAPINI A., *Post-comunicazione. Istituzioni, società e immagine pubblica nell'età delle reti*, Guerini e Associati, Milano, 2014.

PEDUZZI P., *Ma chi ce la fa fare questa Brexit? Storia di una guerra culturale*, in: Il Foglio, 27/05/2017.

PORCARO G. – MULLER H., *Tweeting Brexit: Narrative building and sentiment analysis*, 16/11/2016, in: www.bruegel.org.

PRESINDECIA DE LA REPÙBLICA, REPUBLICA DE COLOMBIA, 2016. *Radican proyecto de ley estatutaria que permite refrendar el día de elecciones posibles acuerdos con la guerrilla.*, in: www.presidencia.gov.co.

REGISTRADURÍA NACIONAL DEL ESTADO CIVIL (Colombia), 2016, in: www.plebiscito.registraduria.gov.

LA REPUBBLICA, *Brexit, dopo i risultati in Gb picco di ricerche su Google: "Che cos'è l'Ue?"*, 24/06/2016, in: www.repubblica.it.

RUIZ B., *The Colombian Civil War*, 2001.

RUSSO F., «La notte del Referendum sui Social Media, ecco come è andata», in *Blog InTime*, 5/12/2016, www.franzrusso.it.

SIXTH SENSE, *Big Data e Referendum Costituzionale*, 2/12/2016, in: www.sixth-sense.it.

SALAMANCA GONZÁLEZ R. and MENDOZA R., «Colombia», in: *Accord Insight 3*, 22-27, 2016.

SANTAMBROGIO A., *Introduzione alla sociologia*, 2014.

LA SEMANA, Rincón W., «Redes sociales en el plebiscito: el fenómeno del yo con yo», 10/04/2016, in: www.semana.com.

SHIRKY C., *Uno per uno, tutti per tutti. Il potere di organizzare senza organizzazione*, Codice Edizioni, Torino, 2009.

SIEGEL A., *'Brexit' Data Report, Social Media and Political Participation Lab*, New York University, 2016.

SIMONETTA B.., *Referendum: ecco come si è schierata la Rete in attesa del voto*, 2/12/2016, in: www.infodata.ilsole24ore.com

SORICE M., «Democratic innovations and political communication. Liaisons dangereuses or new opportunity for democratizing democracy?» in: *Comunicazione politica*, Il Mulino, Fascicolo 3, 2016.

STONE J., *Nigel Farage backtracks on Leave campaign's "£350m for the NHS" pledge hours after result*, www.independent.co.uk.

TATE W., «Paramilitaries in Colombia», in: *The Brown Journal of World Affairs*, Vol VIII, 163-175, 2001.

THE OXFORD DICTIONARIES, *The word of the year 2016*, www.en.oxforddictionaries.com

EL TIEMPO, *Usuarios por el 'Sí', los más activos en redes sociales*, 30/09/2016, www.eltiempo.com

EL TIEMPO, *Las claves para entender el plebiscito por la paz*, 24/08/2016, www.eltiempo.com

TURNEY P., «Thumbs Up or Thumbs Down? Semantic Orientation Applied to Unsupervised Classification of Reviews», in: *Proceedings of the Association for Computational Linguistics*. P. 417–424, 2002.

VINER K., *How technology disrupted the truth*, in: www.theguardian.com.

WOLA, «Colombia Peace» in: *Explaining Colombia's Peace Plebiscite*, 26/08/2016, www.colombiapeace.org

WOOLF N., *Obama is worried about fake news on social media – and we should be too*, in: www.theguardian.com.

Workshop n. 3

BIOLOGY
Toward an Alternative Bioethics
Case Study: When does human life begin?

Giacomo Alfiero, Xavier Gutierrez LC

with
Mayra Cuellar, Guglielmo Gattamelata,
Matteo Ievolella, Nicholas Miles,
Mia Mologousis, Paola Perinu,
Sofia Tilotta, Luca Trabalza

Under the supervision
of Prof. Lucas Lucas SJ
Pontifical Gregorian University

Technology is leading us toward situations we would have never imagined before. In fact, recent developments in biotechnology have broken through human biological limits, one of which is the ability to create human life. It is necessary to follow this change with an anthropological reflection. What, then, is the beginning of life? First, there is a physiological aspect to consider: biological happenings that occur in this phase are strictly related to the metaphysical question about beginning. Thus, a consequent ontological discussion is required in order to outline a global approach to understanding the beginning of life, i.e. what a man and person are, respectively. The focus of this paper is on the beginning of human life in both its meaning and its anthropological deepness, which is why assisted reproduction is the example with which we have chosen to work. This is not a mere ideological presumption, and such an interpretation of the beginning of life is not the result of a certain culture, but is just the opposite: society is built in this way because conception has its own rules, being independent from human will. Heterologous insemination, for instance, can lead to several problems that are not just ethical in nature. As an example, to have an unknown parent blocks medical anamnesis and, although unlikely, could even make unaware incest and all its associated procreative issues possible. Beyond any biased rhetoric, engagement in procreation calls attention to the individual rights of the potentially-resulting person, placing a question mark upon the future of society.

1. Introduction

Our analysis of the beginning of human life in the context of globalization incorporates three different components. The first we will examine is biological in nature, and will require us to explore

the material progress of the human being since the first moment of conception. After careful analysis of this biological component, it will be shown that this being, at the moment of conception, is a new organism of the human species endowed with his parents' genome. Every time there is a new and individual human corporeal material, there is a human being and, therefore, a person, "because that which is biological in man cannot be separated from that which is human." (Lucas 1998: 178)
In a brief philosophical account that acts as this paper's second point of discussion, we will demonstrate that the most modern definitions of the human person do not encompass enough of the human person to serve as a true working definition. The human person is consists of everything that makes it human as well as the totality of its uniqueness; issues of pregnancy hold much weight within the bioethical community because of this continuous struggle to solve the issue of personhood. The global approach of this paper, then, is to underline why issues of biology can also become those of morality. As for the Universal Declaration on Bioethics and Human Rights, the United States represented by O. Carter Snead managed to fight for a change of wording (Cameron 2014: 254), as seen in both Article 3, which now states that "everyone has the right to life, liberty, and security of person" (Snead 2009: 212) and in Article 2, which states that one of the aims of the Declaration is to "promote respect for human dignity and human rights, by ensuring respect for the life of human beings."(Snead 2009: 218) As the universal decree states, let the whole world respect the life of the human being and person.
But then, when does an embryo become a human being? According to some recent judgements, embryos have been considered as "goods", and so subjectable to property-type laws. In fact, in the case *McQueen v. Gadberry*, the Missouri Court of Appeals (n. 103138; Nov. 15th, 2016) was asked to decide if, in the divorce considered, two cryopreserved pre-embryos could be considered as *things* that might come in asset division, or rather as living beings. The court decided that the two pre-embryos could not be considered autonomous assets, and consequently divided them between the two spouses, although they were required to be sub-

ject to a regime of co-ownership as each was inextricably linked to both parties. It is easy to see the huge impact of new technologies in such an interpretation.

It also seems that, because of the individual's status, the embryo is and should be the owner of the fundamental rights granted to all men as such. Among these, as has consistently been affirmed by the ECHR (European Court of Human Rights) itself, there is the right to life, the first and most fundamental presupposition of every other right, and an essential principle of modern democratic societies. In particular, the Strasbourg judges made another important decision regarding the human embryo, recognizing that, already at inception, the embryo cannot be compared to an object and is therefore not patentable, as argued in the case *Oliver Brüstle v. Greenpeace eV* (18/10/2011).

In light of this, it is evident that we need to consider the strong connection that exists between law and technological progress, as the first is often forced to change with the second. Especially when we focus on 'life-beginning technology,' we are called to investigate a universal and unambiguous definition of when exactly life begins, and what the meaning of human really is, and must therefore determine when biology turns into a moral problem. The challenge to define the human being is a biological problem, and once its solution was found then the question of personhood demanded to be answered as well. At this point, biology meets ethics and ethics meet morality; if the human being can exist without the human person existing then certain choices, which exist only as a result of biotechnological development, are ethical and thus moral in nature.

2. Analysis: The beginning of life as a bioethical issue

2.1 From a physiological point of view

The embryonic stage of development is often characterized as the period that lasts until the tenth week of pregnancy,[1] or until the eighth week of development since the first mitotic division of the

[1] "Fetal Development," U.S. National Library of Medicine, February 7, 2017, www.medlineplus.gov

fertilized egg (Findlay et al. 2007) (the first two weeks of 'pregnancy' are considered to start before fertilization, explaining the apparent discrepancy[2]). Throughout this entire work, the focus will thus remain on this stage of development, as it is that which is concerned with inception and is therefore that on which most bioethical debates concentrate.

Additionally, when speaking of the embryo's autonomy it is important to understand that an 'autonomous' being is discussed as possessing a certain independence and ability of control, but not necessarily a freedom from external influence. This is often an area of confusion in embryological bioethics, but is also a topic very central to its understanding, as autonomy is a necessary and sufficient characteristic of the human being.

At the very inception of the embryo, its independence and demonstration of control is immediately evident. In fact, embryonic genome activation (EGA) has been found to occur as early as the 2-cell stage in human embryos,[3] demonstrating the embryo's ability to activate gene expression to continue to foster its growth and organization. During these first few days of cell division, researchers have also noted that there is a sharp decrease in the presence/expression of maternal gene transcripts, (Niakan et al. 2012)[4] illustrating the embryo's growing autonomy as it degrades the genomic instructions provided to it by its mother; instead of using strictly maternal inputs, the embryo looks to sustain its growth via its own, newly-expressed genes as a result of EGA. This EGA leads to the up-regulation of many genes, specifically around the 4- to 8-cell stage, which allows the embryo to continue on its developmental trajectory (*ibidem*).

This gene regulation that is a product of EGA in turn allows for the "remarkable self-organizing capacity of human blastocysts [early embryonic stage]." (Shahbazi et al. 2016)[5] As technology has improved drastically over the past few decades, embryologists finally have the ability to understand the capacity to which the

---

[2] "Fetal Development," U.S. National Library of Medicine.

[3] (Brison D.R., Sturmey R.G. and Leese H.J. 2014).

[4] (Niakan et al. 2012).

[5] (Shahbazi M.N. et al. 2016).

embryo can act autonomously and in a highly-organized fashion, further demonstrating a similarity it shares with the postnatal human. As utilized in an experiment conducted by Javier Herrero, Ph.D., et al., the technique of time-lapsing is being used more frequently given its ease-of-use and its ability to provide details about the embryo's developmental evolution over time.[6] Use of this process has given the scientific community a step-by-step look at the exceptional self-organizing abilities of the human embryo, and the extreme accuracy and predictability with which this organization occurs. From these time-lapse images, one realizes just how precisely the embryo operates, as always establishing its autonomy and control in passing through its various stages as depicted in the above cited experimental report of Kathy K. Niakan et al.

One of the most obvious ways in which one can view the organizational abilities of the human embryo is to examine how its cells act together to prepare for further steps in their developmental process, particularly in relation to their orientation at any given time during embryonic growth. The idea that the orientation of cells within the embryo is orchestrated with an immeasurable precision is evident, for example, when blastocysts attach to only the polar side of the trophectoderm (TE, a particular layer of cells), despite seeming previously disorganized.[7] The same phenomenon is present yet again when the polarization of blastomeres and their position within the embryo occurs,[8] eventually proving necessary for cell determination and differentiation, as will be explained in subsequent paragraphs. The same is true yet again when looking at the process of major embryo reorganization that leads up to gastrulation, the process by which the embryo transforms from a single-layered organism to one that is trilaminar (three-layered).[9]

To attempt to list all of the ways in which the embryo demonstrates its ability to self-organize so efficiently would be impossible and clearly exhaustive, as the supply of examples of this type is virtually infinite. For instance, one could describe the particular way in which TE sub-

[6] (Herrero J. et al. 2016).
[7] (Deglincerti A. et al. 2016).
[8] (Brinson et al. 2014).
[9] (Shahbazi et al. 2016).

types self-organize into concentric rings, or the emergence of discrete territories and lineages of the embryo, both of which are outlined in research conducted by Deglincerti et al.[10] While it is apparent at this point that the human embryo is very autonomous in nature, with the ability to organize and in other ways control various aspects of its own development, not all of the mechanisms by which the embryo does its controlling are understood. Whereas some processes are beginning to be understood, like those by which growth factors are believed to contribute to cell survival and allocation within the embryo,[11] or by which the notochord (a structure providing cartilaginous support) is understood to have an organizational influence on dermal layering,[12] others are harder to comprehend. Regardless of one's ability to understand such occurrences, in many cases it is still easy to tell that there is an extreme coordination by the embryo of the events that take place during its development. For example, before the 4-8 cell stage of the embryo, gap junction intercellular communication (GJIC, as will be discussed later) does not exist. Consequently, one would believe that the cells within this early embryonic stage would not be able to communicate with each other, because science knows of no other means by which the cells could exchange information this early in development. Despite this, human embryos divide synchronously during the first two cleavage events, until GJIC is established, suggesting that there must be another means by which these embryonic cells can communicate.[13] This finding might further propose that even when the exact mechanism by which the embryo acts seem obscure, that does not mean that the embryo is not acting in an orderly fashion and towards an ordered end, continually emphasizing its autonomy.
Of great importance to note is the role of the mother in the organization of the embryo, although her role may be different than anticipated. According to one study conducted by Deglincerti et al.,[14] and reinforced by another conducted by Shahbazi et al.,[15] as

[10] (Deglincerti et al. 2016).
[11] (Niaken et al. 2012).
[12] (Schiebel A.B. 2017).
[13] (Brinson et al. 2014).
[14] (Deglincerti et al. 2016).
[15] (Shahbazi et al. 2016).

long as the human embryo is given an attachment substrate, it is able to self-organize completely independently of maternal influence for up to 12 days after fertilization; yet again, the embryo demonstrates is autonomy and impressive self-organizing ability, performed independently of maternal implantation.

Events directly resulting from the organizational episodes that take place within the growing embryo, determination and differentiation, also demonstrate the embryo's self-sufficiency. It has been reported that determination of cells within an embryo, or their commitment to a certain cell lineage, can occur as early as the 4-cell stage[16] and it even suggested to begin during the pre-patterning event that occurs within the zygote, or newly fertilized egg, before its first 2 divisions.[17] It has been shown over the years that despite an embryonic cell's supposed 'totipotency' or 'pluripotency,' or its potential to form any differentiated cell in the body or most differentiated cells in the body (aside from extraembryoinic cells), respectively, malfunctions in organization or division within the first trimester can have very lasting effects.[18] This supports the idea that embryonic cells do not continue to operate independently of the others, or to retain their toti/pluripotentency, and highlights the way in which they function together as a cohesive unit to establish their unity as a single human being.

Self-organization within the embryo is far more complex than described until this point. While the aforementioned events taking place within the cells no doubt demonstrate their autonomy, what is even more impressive in this respect is the embryo's ability to coordinate complicated processes characteristic of human beings while acting as a singular entity. One of the most striking feats of the human embryo is its beginning to maintain homeostasis, a regulatory process for the upkeep of certain variables within the body. For embryos, this homeostasis is mostly focused on the intra- and intercellular regulation of ions, and clearly expresses the way in which the embryo is capable of controlling its own environment. While relatively little is understood of the embryo's

16 (Niaken et al. 2012).,

17 (Brinson et al. 2014).

18 (Schiebel A.B. 2013).

homeostatic processes, scientific research has revealed its regulation of protons (and thus pH within the cell), as well as other key ions such as calcium, magnesium, and phosphate.[19] In this sense, science's ability to explain the phenomenon of embryonic homeostasis is irrelevant to the idea being examined; all that is needed to understand just how remarkably self-sufficient the embryo is has been already provided by research and experimentation.

Yet another process through which the human embryo displays its autonomy is metabolism. This complex process is a cellular activity beginning even before implantation, as evidenced by experimentation reporting that pyruvate, a key component of aerobic respiration, was readily consumed by the embryo during its preimplantation period.[20] The same research conducted by Leese et al. also points out the increase in glucose consumption during compaction and blastocyst formation, potentially the result of the protein hexokinase's activity within the embryo. As has been the common theme throughout this biological explanation of the embryo's status, this research further stresses the idea that the embryo is independent and, consequently, is able to participate in a large number of tasks distinctive of those cells present within the human adult.

This pattern continues when assessing the embryo's blood circulation and peripheral nervous system activity. By week 3, only one week after implantation, the embryo is developing blood, nerve, and even kidney cells, the heart is already being developed as paired heart tubes, and the brain is shaping into its three main divisions.[21,22] By day 22, the embryo already has a fully-functional and beating heart, and blood vessels continue to disperse themselves throughout the body of the growing embryo, as the brain and its associated neurons continue to expand as well. This clearly proves the embryo's autonomy, as it is able to create for itself functioning organs in a matter of weeks, a process medical researchers have been trying to

[19] (Lane M. and Gardner D.K. 2000).
[20] (Leese H.J. et al. 1993).
[21] (Schiebel A.B. 2013).
[22] "Fetal Development," U.S. National Library of Medicine, February 7, 2017, accessed February 8, 2017, https://medlineplus.gov/ency/article/002398.htm.

replicate for years in order to circumvent organ donation, but one that they have remained relatively unsuccessful at achieving.

The final, and probably most notable, of the autonomous processes carried out by the embryo is that of cell signalling; it is an absolutely imperative process for the embryo because it gives it the ability to foster communication between cells. This makes the organizational processes mentioned previously run more smoothly, and, in light of this, cell signalling is better able to reinforce the idea that the embryo is independent and in-control than is any other aforementioned embryonic operation.

While electrical (neural) communication between cells begins a bit later in the first trimester, mechanical/chemical communication via junctions is believed to be present almost immediately. As previously mentioned, this assessment is based upon data from an experiment conducted by Brinson et al., in which it was seen that despite the nonexistence of GJIC, cells within the embryo still seem to be able to communicate with each other via undiscovered mechanisms given their ability to divide synchronously.[23] Describing what we do know to be true, there are 4 major kinds of junctions that an embryo develops, beginning at the 4-8 cell stage: gap junctions, which are essentially gated channels that directly connect the cytoplasm of two cells and consequently allow them to pass ions, molecules, and impulses from one cell to the next; desmosomes, which hold cells tightly together; tight junctions, which are used to seal adjacent epithelial cells; and adherens junctions, which provide strong mechanical support for attached cells and are believed to be responsible for contact inhibition.[24] Gap and adherens junctions have been proven to form around the 4-8 cell stage, rendering intercellular communication within the embryo "well-established."[25] Additionally, desmosomes and tight junctions appear shortly after this, during blastocyst formation, and continue to provide the embryo with the means to carry-on its growth and development, as well as to demonstrate its autonomy once again. Important to note is also the fact that gap junctions, those open to transfer of ions, etc. between cells, are

[23] (Brinson et al. 2014).
[24] (Alberts B. et al. 2002).
[25] (Brinson et al. 2014).

not passive and exhibit selectivity in which molecules they allow to cross. The embryo, of course, regulates this process of selectivity very carefully, and makes use of phosphorylation/dephosphorylation events, hormones, neurotransmitters, growth factors, and cytokines to further display its control over cell signalling and, more generally, its entire developmental process.[26,27] As "good gap-junctional communication is essential for compaction and embryo development,"[28] the process by which the embryo self-organizes its stages of development is clearly systematic, and therefore ensures the proper functioning of other mechanisms to be utilized later on in its development and the consequent state of unity in which the embryo exists.

2.2 From a philosophical point of view

During the biological development of the zygote, no mutations of quality are made as this would be impossible. This is a human body and the beginning of a human corporeality, and the two ideas cannot be separated. Misconceptions of modern anthropologies has led to a confusing dualism where one must identify the moment in which the spirit becomes a part of the human body. A person is not conceived as an artificial mix of heterogenic components, but as one unified and unique reality, and is an Incarnate Spirit[29]. The body is not only something that we possess. We are our bodies and we live our bodies' life. The human body transcends the sphere of pure belonging for that of being[30] and, therefore, if there is a human body, there is a human being.

2.2.1 Some key distinctions

We have already presented our three-part approach to this question of bioethics. The first two parts may be easy resolvable with biochemical facts, whereas the third requires that we pass from

[26] (Brinson et al. 2014).
[27] (Houghton F.D. 2005).
[28] (Houghton F.D. 2005).
[29] (Lucas Lucas R. 1997).
[30] (Prini P. 1991).

empirical science to metaphysical philosophy. For this passage, we will introduce three key distinctions that will allow us to clarify the anthropological status of the human embryo, specifically whether it should be considered a person or a person "in potency."
The first distinction to be made is that between primary and secondary acts. Primary acts are those acts of being that constitute us as individuals, those that correspond to our human nature. The secondary acts are those in which we exercise our fundamental dimensions or capacities.[31]
The second distinction to be made is that between potential and possibility. Some say that the embryo is a *potential* human being, but they are equating potential and possible. They conceive the embryo as "a collection of cells which, unless it is implanted in the ambience of a human uterus, has no potential of development."[32] We jump to this conclusion only if we have misunderstood the Aristotelian text.[33]
The last distinction to be made is that between complete and concluded. The human embryo is, as we have seen, complete; it is intrinsically constitutes an individual of the human species. It has not, however, achieved the conclusive state of realization, as its essential capacities require biological and psychological maturation, processes that do not end with intrauterine life.
The human embryo can be considered to possess potential with respect to its ability to exercise rationality but we are not dealing with a possible human person, but one that is actual and must integrally develop.

2.2.2 The concept of person

A man has a human nature, something dynamic, rational and intrinsic to the subject that is its essence at an operative level. He is not a cat or a tree but, as said, an embodied spirit. The openness that offers rationality allows us to present the human nature not only as a *factum esse*, but also as a *faciendum*.

---

[31] (Aristotle, *Metaphysics*, IX, 3, 1046b 29).
[32] (*Report of the committee of inquire into Human Fertilisation Embryology* 1981).
[33] (Berti E. 1993: 115-123).

One calls the essence common to all men the 'human nature' and the concrete individual of this human nature a 'person.' So, what is a human person? It is this man. But then one must ask what a person is, and he/she will come to find that it is the individual human being truly existing. The philosophical definition of person is nothing other than the logical expression of the ontological reality of the real human individual; this man is "man" because he possesses human nature.[34]

The concept of person has always been important because it has an ethical and juridical charge. Almost every bioethical debate implies the concept of personhood.[35] This subject is not only the focus of an intense debate but also of great consequences as it determines the extension of our must fundamental rights such as education, justice, welfare and even life.

This debate is why, in 1993, the Bioethics Program of UNESCO was created. This program was broken into two divisions that served as advising boards to the program: the IBC, or the International Bioethics Committee; and the IGBC, which is the Intergovernmental Bioethics Committee. These two distinct parties were intended to collaborate in order to establish universal declarations that UNESCO could potentially incorporate into its governing body. Considering the amount of importance that UNESCO holds within the United Nations, one would assume that this governing body would act efficiently and diligently, and this corporation met his expectation. By October 2003, after only a decade of existence, the unit decided that the time was "'opportune and desirable to set universal standards in the field of bioethics with due regard for human dignity and human rights and freedoms.'" [36] Accepting this major challenge to form a universal standard for the entirety of humanity and its progeny is no small undertaking, so it was no surprise that the assembly proposing the idea, the General Conference (GC), was not entirely sure of what it desired to place within such a document. Therefore, in January of 2004 the GC sent out a simple questionnaire to every country

34 (Lucas Lucas R. 1998: 192).
35 (Macklin R. 1983: 37).
36 (Snead O.C. 2009: 207).

involved in the member-states of the Secretariat of UNESCO, totalling 191 in all. This questionnaire was exceedingly simple, only allowing for a yes or no answer, and the results were fairly surprising; out of all of the potential questions, including those regarding whether or not equality, justice, non-discrimination, honesty, or free consent should be addressed within the proposed document, the only issue that one hundred percent of the polled member-states agreed should be included in the document was respect for human dignity.[37]

Upon creation, the document was named the Universal Declaration on Bioethics and Human Rights, and contained several provisions and clauses that seek to protect the human person. Through this paper, UNESCO hoped to solve most bioethical issues and debates. However, modern bioethics has become a major area of contention due to the aforementioned concept of the person. This debate of personhood has plagued the world since the beginning of philosophy, and no clear and distinct answer is yet to be found. Unfortunately, the debate is often fuelled by political, social, or partisan attitudes which will be explained through example using parts of the Declaration.[38] The central ideas of this portion, however, is to discover whether or not one can determine an adequate definition for the human person, why this definition is so pertinent to conversations of embryonic status, and where the problem of bioethics transforms from a biological discussion into an ethical and moral one.

The definition of the human person that is most widely accepted is that crafted by John Locke, who holds a reductionist viewpoint that finds the human person solely within the consciousness. In fact, in Locke's mind, the person must be a thinking substance that can ascertain itself. This theory was subsequent to the definition that Descartes himself gave in his *Second Meditation* where he stated that one's identity could only be found within one's thinking of self. Therefore the self, to Descartes, needed to be aware of its own existence in order to be an identity. Locke agrees with Descartes on this point, but goes further in saying that the identi-

[37] Idem, 208

[38] (Cherry M.J. 2009: 285).

ty of person exists solely in the mind as well.[39] Notice that it is not the identity of the person that is found, but in fact personhood itself and, therefore, personhood can only exist if the said person is conscious of its existence. To this effect, Locke even says that multiple persons could inhabit the same body, similar to the idea of multiple personality disorder.[40]

This definition provided by Locke is not the only one that is accepted by bioethicists. Boethius claimed that the human person was "an individual substance of a rational nature."[41] This definition is widely accepted as well, but may leave out those who are in a persistent vegetative state (PVS).[42] A.A. Howsepian concluded that consciousness is not essential to personhood at all, saying that just having human DNA or a neocortex would make one a person. Unfortunately, this argument seems to justify calling individual cells of skin persons, as they do hold the entirety of one's DNA, despite most disagreeing with this idea.[43] Others, like the bioethicist Walter Glannon, claim the opposite: that cognitive ability is absolutely essential to personhood. Glannon adheres to this belief so strongly that she claims that both PVS patients and the profoundly disabled should not be considered persons. Robert Troug takes this idea so far as to write that we should harvest organs from living PVS patients.[44] Clearly, there are several differing viewpoints that are present within the bioethical field.

Returning to Locke then, what issues could one find with his definition of personhood? First, Locke's reductionist theory does exactly as its name suggest: it reduces the human person to the consciousness. Therefore, one's experiences make up one's person, which is not meant to say that Locke believes that one's actions make the person, but rather that the person must be consciously aware of the experiences that affect it either intrinsically or extrinsically. Here the reader may begin to see the flaw in this idea and

[39] (Meador, Keith G., and Shuman J.J. 2000: 77).
[40] (Crosby J.F. 2000: 38).
[41] (Smith, Allyne L. Jr. 2000: 3).
[42] Idem, 3
[43] (Howsepian A.A. 2000: 17).
[44] (Smith, Allyne L. Jr. 2000: 5-9).

the beginning of how the issue of personhood becomes an issue of morality. In Socrates' *Gorgias*, he states that it would be better for a person to physically suffer than to inflict self-harm upon one's soul by committing a moral injustice. Not assuming that Socrates would believe that soul and person are synonymous, it is plausible to transfer this same meaning to personhood as well. Extrapolating then, it would be better for a person to physically suffer than to harm their personhood through moral injustice because ethics are generally prescribed to good people.[45] However, one does not experience harm from committing some morally wrong act aside from a physical harm like punishment or guilt. Since this harm is not experienced then, it could not belong to the human person according to Locke. To whom does the harm belong? It is not possible to ascribe this action to all of humanity or even to anyone other than the person that committed the act. Some may argue that actions similar to this would be taking away from the personhood of the actor, but even then this effect is not experienced. Therefore, it must be accepted that there are things outside of experience that affect an individual person, and Locke's argument is proven to be insufficient.

After all of this debate there has not been a single definition that has ascribed to a proper view of person, and so one must be provided that allows the best possible answer. As to this conundrum Orthodox Metropolitan John Zizioulas of Pergamon may assist with an answer. In 1991, Zizioulis wrote,

> Personhood is not about qualities or capacities of any kind: biological, social or moral. Personhood is about hypostasis, i.e. the claim to *uniqueness* in the absolute sense of the term, and this cannot be guaranteed by reference to sex or function or role, or even cultivated consciousness of the 'self' and its psychological experiences, since all of these can be classified, thus representing qualities shared by more than one being and not pointing to absolute uniqueness.[46]

[45] (Crosby J. F. 2000: 44).
[46] (Zizioulas J. and McPartlan P. 2006).

This definition is supplemented by the necessity of persons being bodily creatures as well as having this uniqueness that does not belong to any bodily quality. Therefore, the person must be made up of both the mentioned hypostasis of being and the physical body, and does not exist without this union.[47] However, this hypostasis cannot simply refer to animals. This uniqueness includes what makes humans distinctly human, mainly free will and the ability to reason, as well as the entirety of that individual to form a human person.

Perhaps this definition could better answer the questions that we have about what makes a person a person, but then it leaves one to wonder about when the person can come into existence. Therefore, pregnancy becomes a vital topic of interest in bioethics and personhood. Bioethicists are no longer in contention about when a human is formed, as is demonstrated by the statement: "unequivocally, human life begins at conception."[48] To further prove this point, one can again look back to the formation of the Declaration on Bioethics and Human Rights. After this document was submitted to 'expert notation' stage, some countries specifically attempted to substitute the word "person" for "human being" throughout the entire documentation. The four countries that heralded this change were the Netherlands, Japan, the United Kingdom, and South Korea. These four countries desired this change in order to protect the destruction of embryos for the sake of research as well as for abortion laws.[49] Clearly, the human life begins at conception. However, there are those that argue that the human person does not start as the human being, and this is why bioethics frequently returns to issues on abortion.

As a definition of person has already been provided that would allow for human personhood to exist at the point of conception, possibly viewing another religion's opinion on the matter could provide better insight. In Islam, the world's second largest religion behind only Christianity, the human person begins 120 days after conception through a process called "ensoulment." This process is

[47] (Crosby J.F. 2000: 47).

[48] (Smith, Allyne L. 2000: 7).

[49] (Snead O.C. 2009: 214).

referred to in the genesis story of the Quran in which God created man for three sections of forty days and then blew His spirit into man. However, in Islam the fetus is afforded the same rights as any other human person. The fetus has the right to life, inheritance, and the punishment of anyone who attempts to harm it. Also, it is immoral to discard the fetus for any reason. If the fetus is miscarried, one should use the fetal tissue for research or to attempt to conceive again. Human cloning is outlawed for the same reason: God creates every human *uniquely*.[50] The justification used to protect the personhood of embryos in Islam is the same as that which was already mentioned, the person holds its own unique humanity.

Christian thinking also supports the idea that personhood begins with conception. Since most people object to this idea by stating Locke's reasoning, which was already proved insufficient, one can again answer this reasoning with strictly Christian thinking. If consciousness is necessary for personhood, then what do the sacraments affect? Since one cannot experience a specific sacramental grace, which remains a mystery to humanity, then it should not be affecting a person in Locke's point of view. However, the sacraments were created to affect the human person through the soul, and not merely for the soul's sake. Therefore, Locke's argument again does not hold. Since a Christian can practice infant Baptism, the Christian must believe the baby is a person despite being incapable of self-awareness and experience until the age of eighteen months.

3. Problems: how do we deal today with the beginning of persons?

Biotechnology is a set of technologies that involves living beings carrying out their own biological processes to obtain/modify products of interest. It can now be applied to areas such as medicine/health, agriculture/food, and zootechnics, or, most revolutionarily, to the human being (Sorge 2011: 369). This application is not immune to controversy, however, as it may cause the ma-

---

[50] (Athar S. 2008: 61-62).

nipulation or even destruction of the life itself (Lucas 2001: 11). We live in a society where "patents on life strike at the heart of our convictions about the true nature of life, that is, whether it should be considered as an intrinsic value or a simple utility value" (Rifkin 1998: 113). In his encyclical *Caritas in Veritate*, Pope Benedict XVI himself highlights the risks of this new technology that is imposed as a dominant single thought in a globalized world (2009: 70).

Biotechnology forces an anthropological question about the manner in which a human life is conceived. With this new technology, human life can be manipulated in many ways, from *in vitro* fertilization to embryonic research, cloning and human hybridization (Sorge 2011: 96). The human *embriopoiesi* is the process by which, through intervening reproductive technologies, the human embryo develops and is monitored (Russo1994: 17). The point is to define the moment at which the embryo begins to exists as an individual called 'man.' Understanding of this question is crucial in order to assess when embryonic interventional techniques are morally permissible, and when they are not.

3.1 The impact of new technologies on the beginning of human life

Technology has been drastically increasing its role in the medical field over the past years. There are many new techniques that could be used for helping the unborn or mother both before and during gestation, such as artificial procreation, embryonic manipulation or experimentation, but not all of these techniques can be deemed morally acceptable. In order to determine which are morally acceptable and which are not, we must focus on the role of and the dignity bestowed upon the human embryo in society. This society has asked of the law: when does an embryo have the same rights as a person and, similarly but separately, when is it considered 'human'? Traditionally the law has considered all humans as human regardless of generation mode and has set forth rules that aim to protect this human life from acts such as abortion and murder that speak to a human conduct that is considered equally as detrimental to the human being, just at different periods of its existence.

In many legal systems throughout the world, rules have been introduced that punish or do not permit abortion outside of certain circumstances,[51] but there remain serious gaps in legislation about the explicit rights of the unborn. This is true not only in many state systems but also in international declarations such as those from the European Parliament (on genetic manipulation and in vitro fertilization), the European Council (on utilization of embryos and fetuses in the scientific research), and the American Declaration of the Rights and Duties of Man[52].

The use and role of the embryo has been very controversial in jurisprudence. For example, Article 6 of a directive of the European Parliament (98/44/CE) declares "inventions shall be considered unpatentable where their commercial exploitation would be contrary to public order or morality; however, exploitation shall not be deemed to be so contrary merely because it is prohibited by law or regulation," and that the "use of human embryos for industrial or commercial purposes" is prohibited. This directive requires that the national court play a fundamental role in determining, in light of scientific developments, whether a stem cell obtained from a human embryo at the blastocyst stage constitutes a 'human embryo' within the meaning of Article 6 (no. 2, letter C) of Directive 98/44. In addition to that, the German Federal Court of Justice declared the invalidity of a patent that was already granted by the European Patent Office on progenitor stem cells (cells with differentiative abilities only slightly less than those of stem cells) extracted from human embryos on the basis of national patent law, patentgesetz, that reversed Article 6 (2c) of the European Directive EC / 98/44 which prohibits the patenting of human embryos.

A legal definition of an embryo is can be read from the *Embryonenschutzgesetz* (ESchG, Embryo Protection Act, 1990. Trad. BGBl. I S. 2746) where it says under paragraph 8: *an embryo is a fertilised human ovum capable of development, from the time of karyogamy, and any cell removed from an embryo which is 'totipotent', that is to*

[51] The norm reproduces Article 1 of the antecedent French law of 1975.

[52] Also bear in mind the "European Statement on bioethics" from the European Council. On the entire problem and for an accurate documentation. (Balestrero E. 1997).

say, able to divide and develop into an individual provided that the other conditions necessary are satisfied. A distinction must be made between those cells and pluripotent cells, which are stem cells which, although capable of developing into any type of cell, cannot develop into a complete individual.

On the case of *Oliver Brüstle v. Greenpeace eV* (18/10/2011), the European Court of Justice stated that the term embryo is to be given a broad meaning and concluded that, at the moment when a human egg is fertilized, it must be considered embryo. In addition, the Court ruled that the term embryo must include every human unfertilized egg whose nucleus has been transplanted into a mature human cell, as well as a human unfertilized egg whose division and further development have been stimulated by parthenogenesis. The Court also stated that "any human ovum must, as soon as fertilised, be regarded as a 'human embryo' within the meaning and for the purposes of the application of Article 6 (2c) of the Directive, since that fertilisation is such as to commence the process of development of a human being. That classification must also apply to a non-fertilised human ovum into which the cell nucleus from a mature human cell has been transplanted and a non-fertilised human ovum whose division and further development have been stimulated by parthenogenesis."

Technology plays an especially delicate role in relation to fertility, maternity and the so-called 'right to conceive,' given the outstanding advancements that have been made in the field of assisted reproduction over the past few decades. The debate on assisted reproductive techniques (which include procedures from *in vitro* fertilization to 'surrogate motherhood'[53]), though not a new one, still appears to create factions within society, namely between positions rooted in religious beliefs and those in favor of broadening the scope of, and access to, technological solutions for reproduction. Briefly summarized, the fundamental line that divides the aforementioned factions stems from the moral/ethical dilemma that is at the heart of the issue being considered, due to the implications it carries regarding man's ability to impact life.

[53] See http://www.thehastingscenter.org/briefingbook/assisted-reproduction/

Understanding that line is a complicated task, but there are both theoretical and practical tools that can offer guidance and allow one to at least attain an educated and critical consciousness on a topic which even lawmakers have, to this date, preferred to refrain from defining and regulating in a straightforward manner.

Pope John Paul II outlined the issue when he reminded us of the "ambivalence" of technological advancements today, which enable man "to take in hand his own destiny, but [which] also expose him to the temptation to go beyond the limits of a reasonable dominion over nature," underlining the "implicit dangers" of such potential influence when applied to maternity and birth.[54] A crucial question arises: what might these dangers be, which call for a limitation in the application of reproductive technology? The first step towards an answer appears to be realized in the Compendium of the Social Doctrine of the Church, which states that "the desire to be a mother or a father does not justify any 'right to children,' whereas the rights of the unborn child are evident. The unborn child must be guaranteed the best possible conditions of existence through the stability of a family founded on marriage, through the complementarities of the two persons, father and mother."[55]

There thus appears to be at least two basic elements that come into play: on one hand, the necessary connection between conception and the conjugal act and, on the other, the priority of protecting the unborn child's rights above the desire to have a child. The first aspect implies that maternity and paternity be the consequence of a natural interaction between a man and a woman, generally bound by wedlock. It is not merely a matter of dogmatic adherence to a set of principles that might be considered to belong mainly to the religious domain; assisted reproductive techniques, even when they are aimed at circumventing infertility and do not involve third parties, still carry significant issues on the treatment,

[54] (John Paul II, 1980). Discourse to those taking part in the 81st Congress of the Italian Society of Internal Medicine and the 82nd Congress of the Italian Society of General Surgery, 27 October 1980. https://w2.vatican.va/content/john-paul-ii/it/speeches/1980/october/documents/hf_jp-ii_spe_19801027_congressi-medicina.html

[55] Compendium of the Sociale Doctrine of the Church. http://www.vatican.va/roman_curia/pontifical_councils/justpeace/documents/rc_pc_justpeace_doc_20060526_compendio-dott-soc_en.html

conservation and, most sensitively, the acceptability of even discarding human embryos.[56] The issue of whether an embryo deserves the same rights and protection of a human being is far from resolved, while the lack of clear legislation or regulatory guidelines on these matters leaves a grey area in which questionable private practices are allowed to develop.

More specifically, there is a generally recognized risk of commodification of the unborn human being as it becomes the object of a trade in which its rights are devoid of protection. This gives way to the fulfillment of the desires of individuals and couples determined to have a child, while access to this same market is limited by financial discrimination.

These critical objections appear to be challenged by those who support a stronger affirmation or recognition of the 'right to have a child,' which includes promoting greater access to assisted reproductive techniques. Evidence of this approach can be found both on a national level and within the international community. The former, could be argued, is true for Italy, where potential limitations introduced by lawmakers on certain technological procedures in assisted reproduction were progressively overruled by the judiciary.[57] Similar notions in favor of greater access to reproductive techniques appear to be proclaimed in a series of international documents, starting from the 1948 Universal Declaration of Human Rights, in which article 15 explains the "right to form a family," a broad principle that, applied to today's reality, lends itself to a variety of interpretations, such as assisted reproduction.[58] Nonetheless, the two connected issues of commodification and financial discrimination do not at this point appear to have been solved even by an increasing emphasis on a right to have a child, a conclusion that is not just supported by theoretical reasons previously discussed, but that is also suggested by practice and especially when third-party assisted reproduction is being considered. Several countries legally recognize the practice of surrogacy,

---

[56] (Brezina P., Zhao Y. 2012).

[57] See http://www.jurist.org/paperchase/2014/04/italy-constitutional-court-overturns-assisted-reproduction-restrictions.php

[58] (Schenker J. 2011).

whereby a woman is bound by contract to give birth to a child which is then delivered to another party (generally a heterosexual couple, according to local legislation) in return for payment. A 'human market' that is known to be especially blossoming in less-developed countries where low standards of living, failing social security and widespread social imbalances often leaving single women without support combine with easy access to technology and low costs, attracting what has come to be known as "reproductive (or fertility) tourism."[59]

Assisted reproduction techniques indeed are aimed at achieving artificial conception, both by intracorporeal and extracorporeal means. It is sometimes emphasized assisted reproduction therapeutic purposes already involved in the different levels of biological mechanisms failure. In some cases, however, this is not just a help or care of a therapy; it seems nothing but a desire.

These techniques have developed as appropriate solutions to the problem of infertility. Even on this subject there is a distinction: the causes of infertility are not only physical, as in that which is due to the reproductive functioning of the parents, but also moral as it relates to choices made by the parties involved. For instance, a woman's fertility at an advanced age can be compromised by a decrease in the quantity of potentially fully-functioning oocytes she possesses, and/or by previous sexual behavior that has perhaps made her sterile due to infection (Bompiani, 2006: 107). Thus, techniques for assisted reproduction are strictly related to a worldview, and life, as well as the concepts of motherhood and fatherhood, is seen mostly as a right and not as a gift (Russian, 1994: 31).

We must separately evaluate each of these techniques, taking into account the human values involved.

GIFT, the Gamete Intra-Fallopian Transfer, is an artificial intracorporeal fertilization technique that involves inserting oocytes into the fallopian tube, together with treated sperm. Providing an advantage over other types of in vitro fertilization (IVF), this

[59] See also https://graceuniversity.edu/iip/2014/11/ethical-issues-gestational-surrogacy/ and http://espresso.repubblica.it/inchieste/2017/01/05/news/noi-donne-ucraine-in-offerta-come-mamme-surrogate-1.292921

GIFT technique does not involve the direct manipulation of the embryo, but does present the risk of miscarriage.
Instead, the IVF embryo transfer is a different reproductive technology in which fertilization occurs in a production plant and guarantees a greater chance of success. The oocytes are put in contact with the sperm, using IVF and the resulting embryos are placed in the uterus about three days later. In fact, the IVF method requires the availability of fertilized embryos, and allows for the possibility to repeat the implantation process in case of initial failure.
The method, then, has several ethical implications at the scientific, social and genetic levels.
At the socio-ethical level, there is a distinction of procreation in the intimately-linked sexual act of the couple and sexuality-procreation dissociation. With IVF, the gametes may not belong to the couple making use of them. The resulting child can be brought up by legal parents but can be a genetic stranger to at least one of them. There are other different issues to consider with respect to this technique, e.g. the deliberate loss of scheduled early embryos, so-called residual embryos, the use of the embryo for research or experimentation, risk of embryonic and consequent postnatal malformations, and the risk of abortion of the embryo both before and after the transfer of the uterus. We can therefore say that it "realizes a birth that is not worthy of the human person, because it does not guarantee a real interpersonal relationship in the man-woman-triad generated, which is the basis of being born" (Russian, 1994: 40). At the social level, we can talk about genetic manipulation resulting in the potential for designer babies, or those whose traits could one day be hand-picked by parents, for instance to determine the sex or eye color of the child, etc. This technique gives no regard to the psychological development of the child due to a growth environment may not be ideal.
The case of surrogate maternity is peculiar, and is very stressed in recent times. In such a context the damage on a resulting child's development can be substantial; indeed, he receives the genetic heritage of two people, the vital intrauterine connection by third (the surrogate mother), to finally be raised and educated by the

legal parents. The child has the right to be conceived according to the truth of birth and his being a person, as has been previously mentioned. Thus, he has the right to grow consequently to his conception, and according to a unique interpersonal relationship and unity with the mother: "no one may be able to affect the deep interactions, especially on a psychic level, that are established between the baby and its mother" (Rifkin: p. 44). All of these new biomedical practices allow for a widespread use of embryos and are called therapeutic in that they make use of cells and tissues in dead embryos for treatment of certain diseases or for research purposes to observe the first stages of embryonic development in intra-corporeal fertilization. The act of keeping alive human embryos in vitro for experimental purposes demonstrates abuse of personal dignity, and beyond this there is even a commercial issue to consider. As mentioned by Rifkin, "genes are the green gold of the century of biotechnology," and there is even, for example, a market for the sale of dead embryos to be used in cosmetic manufacturing (Rifkin: 77).

The reasons for using biotechnology for sex determination are several, from the prevention of genetic diseases related to sex to couples looking for a child of wanted sex out of a simple desire or an obligation to give birth to a sex deemed superior. This process involves manipulation of the female reproductive tract for the creation of certain mechanical and biochemical conditions that allow for greater viability of sperm types, IVF, the sperm separation by centrifugation and other techniques.

In short, biotechnology represents hopes and aspirations, but also fears and suspicions. They aim at the heart of one's identity, represent the man's control of nature and provide an affordable option to conceive, but at what price? The availability of the embryo seems to depend on the will of others, but is it to be considered as a human person or as a tool that can be used?

This question, in conclusion, lends itself to Pope John Paul II's words on the "implicit dangers" of technology applied to life when human desires, or financial means, surpass the rights of those unborn, thus warranting their greater protection.

4. Conclusions

Human needs often resulting from reproductive failures lead to the exasperations of making the reproductive process become a productive one. These failures are, most commonly, sterility and the suffering of serious pathologies, issues for which the remedy has been identified in assisted reproduction. In what follows, the aim is to put aside all risks and liabilities connected to assisted reproduction but to focus on its moral legitimacy through an anthropological-ethical approach.

Given that human dignity shall be the basis from which to understand and to evaluate what is happening, the starting point must therefore be determining the moment at which human life begins to exist. This question is clearly not a trivial matter, and is necessary to answer because of the change in life's origin that occurs during the assisted reproduction since it provides a new means of procreation.

Human life follows the willingness of a man and a woman to bring themselves together and so impress their genetic and personal characteristics upon this new form of life to be called their child. This genetic impression occurs at the embryonic stage, and so, as has been demonstrated, personal identity does exist in this new individual, this new human being.

The above claim, however, is still not accepted by all, as many believe that in so far as the embryo does not have the typical external appearance of a human being, it cannot be considered equivalent to the latter. This proves to be a mistake, though, as it is based on a vision which reduces human beings to their corporeity, or their external appearance. While assessing the validity of this claim, it is necessary to ask how the embryo could become a human being if not yet from its existence, a question difficult if not impossible to answer.

Hence, it has to be pointed out that from an anthropological and biological perspective what has been referred to as zygote, blastocyst and so on is the beginning of a new human being, which has already had in itself that body that it will continue to transform during the phase of the life-cycle that ends with birth.

As the Assembly of the Council of Europe observed, "the human embryo, though displaying successive phases in its development which are designated by different terms (zygote, morula, blastula, pre-implantation embryo or pre-embryo, embryo, foetus), displays also a progressive differentiation as an organism and nonetheless maintains a continuous biological and genetic identity".[60]

The development of the embryo happens according to and because of a substance intrinsic to itself, which brings with it all these features that are going to be shown on the adult body, so that it (i.e. the embryo) shall have a human dimension. Thus, having acknowledged that the embryo is a human being at the moment of conception, the moral duty to legally protect it is apparent.

As already pointed out, human life results from of willingness of two parties, so in such a union both an affective and a procreative dimension (co)exist.[61] Given this, assisted procreation almost lacks one of the aforementioned dimensions: the procreative one. [62] It implies that assisted procreation, in any way performed, can never be accepted because it leads to human life coming into existence outside of the woman's body. Therefore, it exposes the embryo, a human being worthy of respect and protection, to manipulation and experimentation.[63]

---

[60] *Use of human embryos and foetuses in scientific research,* Parliamentary Assembly of the Council of Europe, Debate on Feb. 2nd, 1989 (https://assembly.coe.int/nw/xml/XRef/Xref-XML2HTML-en.asp?fileid=15134&lang=en).

[61] «L'origine dell'essere umano risulta così da una procreazione legata all'unione non soltanto biologica ma anche spirituale dei genitori uniti nel vincolo del matrimonio» (Congregazione per la Dottrina della Fede 1987).

[62] The term "almost" has been used because namely considering the heterologous in vitro fertilisation, it also lacks the affective dimension.

[63] Assisted procreation cannot be justified even on the reason of sterility or inherited pathologic conditions. In fact, if the latter are assumed to be irreversible situations deserving to be solved by medical science and technology, it should be likely assumed that assisted procreation is a therapy. But it is not, for the simple reason that once it is performed, sterility or any pathologic illness is not resolved. Consequently, assisted procreation may be seen just as a medical technique through which a personal act (procreation in itself) becomes an impersonal mechanism leading to the final outcome of having a child, but not to create a human being.

The deviation of the assisted procreation from the natural conception process appears even more clear from a practical point of view. The first critical issue lies in the fact that, in what is expected to be a new way to procreate, there are several players and roles, with a consequent depersonalization of the generative aspect of life. There is a couple wishing to have a child, as well as healthcare professionals working within a specific legal framework toward the desired product. In this scenario, there is not any space for 'life,' because a child is just the outcome.

References

AA.VV., *La vida humana: origen y desarrollo*, Federación internacional de Universidades Católicas, Madrid, Barcelona, 1989.

AGAZZI E. (ed.), *Bioetica e persona*, Milano, Angeli, 1993.

ALBERTS B. – JOHNSON A. – LEWIS J. – RAFF M. – ROBERTS K. – WALTER P.. *Molecular Biology of the Cell*, 4th ed., Garland Science, New York, 2002.

ATHAR S., «Enhancement Technologies and the Person: An Islamic View.» in: *The Journal of Law, Medicine & Ethics* 36.1, 59-64, 2008.

BENEDICT XVI, *Caritas in veritatis*, in: w2.vatican.va/content/benedict-xvi/it/encyclicals/documents/hf_ben-xvi_enc_20090629_caritas-in-veritate.html

BIOLO S. (ed.), *Nascita e morte dell'uomo. Problemi filosofici e scientifici della bioetica*, Marietti, Genova, 1993.

BOLE T. J. III, «The Person in Secular and in Orthodox-Catholic Bioethics.» in: *Christian Bioethics* 6.1, 85-112, 2000.

BOMPIANI A., *Le tecniche di fecondazione assistita: una rassegna critica*. Vita e pensiero, Milano, 2006.

BRISON D.R. – STURMEY R.G. – LEESE H.J., «Metabolic Heterogeneity During Preimplantation Development: The Missing Link?» in: *Human Reproduction Update* 20, no. 5, 632–40, May 2, 2014, doi:10.1093/humupd/dmu018.

CAMERON N.M. De S., «Humans, Rights, and Twenty-first Century Technologies: The Making of the Universal Declaration on Bioethics and Human Rights.» in: *Journal of Legal Medicine* 35.2, 235-72, 2014.

CHERRY M.J. UNESCO, «Universal Bioethics and State Regulation of Health Risks: A Philosophical Critique.» in: *Journal of Medicine and Philosophy* 34.3, 274-95, 2009.

COMITATO NAZIONALE PER LA BIOETICA, *Identità e statuto dell'embrione umano*, Presidenza del Consiglio dei Ministri, Roma, 1998.

CONGREGAZIONE PER LA DOTTRINA DELLA FEDE, *Donum vitae*, LEV, 1987.

CROSBY J.F., «Person and Consciousness.» in: *Christian Bioethics* 6.1, 37-48, 2000.

DEGLINCERTI A. – CROFT G.F. – PIETILA L.N. – ZERNICKA-GOETZ M. – SIGGIA E.D. – BRIVANLOU A.H., «Self-Organization of the in Vitro Attached Human Embryo.», in: *Nature* 533, 251–63 , 2016.

DUSKA R., «On confusing human beings and persons», in: *Realism*, edited by D.O. DAHLSTROM D.O., 158-165. American Chatolic Philosophical Association, Whasington, D.C., 1984.

ELFSTROM G., «Are chimpanzees persons?» in: *Journal of the Alabama Academy of Science*, vol. 86, no. 2, p. 78., 2015.

ENGELHARDT H.T., *Manuale di bioetica*, Il Saggiatore, Milano, 1999.

FINDLAY J.K. – GEAR M.L. – ILLINGWORTH P.J. – JUNK S.M. – KAY G. – MACKERRAS A.H. – POPE A. – ROTHENFLUH H.S. – WILTON L.. «Human Embryo: A Biological Definition.» in: *Human Reproduction* 22, no. 4, 905–11, 2007.

FORD N.M., *Quando comincio io? Il concepimento nella storia, nella filosofia e nella scienza*, Baldini & Castoldi, Milano, 1997.

GIANNARAS C., *Elements of Faith: An Introduction to Orthodox Theology*, T & T Clark, Edinburgh, 1991.

HERRERO J. – TEJERA A. – ALBERT C. – VIDAL C. – DE LOS SANTOS M.J. – MESEGUER M., «A Time to Look Back: Analysis of Morphokinetic Characteristics of Human Embryo Development.» in: *Fertility and Sterility* 100, no. 6, 1602–9, 2013.

HOUGHTON F.D. «Role of Gap Junctions During Early Embryo Development.» in: *Reproduction* 129, no. 2, 129–35, 2005.

HOWSEPIAN A.A., «Toward a General Theory of Persons.» in: *Christian Bioethics* 6.1, 15-35, 2000.

IDE P., *Le zygote est-il une personne humaine?*, Téqui, Paris, 2004.

IPPOLITO S. – CIRELLI C. (cur.), *I fondamenti della dignità della persona*, Loffredo, Napoli, 2006.

LANE M. – GARDNER D.K., «Regulation of Ionic Homeostasis by Mammalian Embryos.» in: *Seminars in Reproductive Medicine* 18, no. 02, 195–204, 2000.

LEESE H.J. – CONAGHAN J. – MARTIN K.L. – HARDY K., «Early Human Embryo Metabolism.» in: *BioEssays* 15, no. 4, 259–64, 1993.

LUCAS LUCAS R., (cur.), *Commento interdisciplinare all'Evangelium Vitae,* LEV, Vaticano, 1997.

— *Antropologia e problemi bioetici*, Edizioni San Paolo, Milano, 2001.

MEADOR K.G. – SHUMAN J.J., «Who/se We Are: Baptism as Personhood.» in: *Christian Bioethics: Non-Ecumenical Studies in Medical Morality* 6.1, 71-83, 2000.

NIAKAN K.K. – HAN J. – PEDERSEN R.A. – SIMON C. – REIJO-PERA R.A., «Human Pre-Implantation Embryo Development.» in: *Development* 139, no. 5, 829–41, 2012.

PALAZZANI L., *Il concetto di persona tra bioetica e diritto,* Giappichelli Editore, Torino, 1996.

PERRING C., «Against scientism, for personhood» in *The American Journal of bioethics*, January (1) 2007.

PONTIFICIA ACADEMIA PRO VITA. *Identity and Status of the Human Embryo: Proceedings of the Third Assembly of the Pontifical Academy for Life (Vatican City, February 14-16, 1997).*

2nd ed. Edited by Elio Sgreccia and Juan de Dios Vial Correa. Vatican City: Libreria Editrice Vaticana, 1999.

RIFKIN J., *Il secolo bioetico: il commercio genetico e l'inizio di una nuova era*, Baldini&Castoldi, Milano, 1998.

RUSSO G., *La bioetica e le tecnologie della vita umana nascente*, Elle Di Ci, Torino, 1994.

SACKS D.M., «Fetal Development.» in: https://medlineplus.gov/ency/article/002398.htm.

SCHEIBEL A.B. «Embryological Development of the Human Brain.» May 15, 2013. Accessed February 6, 2017. http://education.jhu.edu

SHAHBAZI M.N. – JEDRUSIK A. – VUORISTO S. – RECHER G. – HUPALOWSKA A. – BOLTON V. – FOGARTY N.M.E., Et Al. «Self-Organization of the Human Embryo in the Absence of Maternal Tissues.» in: *Nature Cell Biology*, 18, 700-708, 2016.

SINGER P. – BUCKLE S. – DAWSON K., «The singamy debate: when precisely does a human life begin?» in: *Embryo Experimentation,* Cambridge University Press, Cambridge, 1990.

SLACK J., *From Egg to Embryo; Regional Specification in Early Development*, Cambridge University Press, Cambridge, 1991.

SMITH A.L., Jr. «Personhood: Beginnings and Endings.», in: *Christian Bioethics* 6.1, 3-14, 2000.

SNEAD O.C., «Bioethics and Self-Governance: The Lessons of the Universal Declaration on Bioethics and Human Rights.» in: *Journal of Medicine and Philosophy* 34.3, 204-22, 2009.

SORGE B., *Introduzione alla dottrina sociale della chiesa*, Queriniana, Brescia, 2011.

SPAEMAN R., *Persone. Sulla differenza tra «qualcosa» e «qualcuno»*, Laterza, Roma-Bari, 2005.

VANDRUNEN D., «What Is Christian About Christian Bioethics? A Reformed, Covenantal Proposal.» in: *Christian Bioethics* 21.3, 334-55, 2015.

WATT H., «Potential and the early human» in *Journal of Medical Ethics* 22, 222-226, 1996.

— «The origin of humans» in *Linacre Quarterly*, 62, 36-46, 1995

WHELTON B. J. B., «On the beginning of human life» in *Linacre Quarterly*, 65, 51-65, 1998-1.

YOUNG K., «The zygote, the embryo, and personhood: an attempt at conceptual clarification» in *Ethics Medicine*, 10 (1) 2-7, 1994 Spring.

ZIZIOULAS J. – MCPARTLAN P., *Communion and Otherness: Further Studies in Personhood and the Church*, T & T Clark, London, 2006.

Workshop n. 4

ECONOMICS Changes in the Financial System Case Study: Impact Investing

Mariangela Lancellotta, Adrian Lawrence LC

with
Samuele Crosetti, Marco Matani,
Maria Kunath, Annamaria Roberto,
Arnaldo Mitola, Irene Buzzi,
Andrea Cavicchi,
Constantino N. Christoyannis,
Antonino Cordopatri

Under the supervision
of Prof. Fernando De La Iglesia Viguiristi SJ
Pontifical Gregorian University

1. Introduction

In the last few decades, the financial system has grown thanks to a technological progress, which has tremendously boosted its potentialities. Yet the recent crisis of 2008 has also significantly affected it.

After a brief analysis of the evolution of economic theories, this paper focuses a trend of an alternative finance market able to transform significantly the basic dynamics of the same system, this trend is the so-called "impact investing". This approach stems from the awareness that impact-investing is distinguished as a hybrid and seems to be consistent with the values and investment strategies of the new generations that will lead the future of the market in the coming years.

Today, impact investing is still shaping and developing. In the global market, every country is trying to mark criteria for investment choices, targets, expected outcomes and objectives that identify them. Italy is also taking the first steps and is positioning itself in the European landscape as an important player, even at a legislative level. In addition, the Vatican also has significantly drawn attention to this issue. The authors have chosen to leave out the topic of impact measurement, considering it as not falling within the scope of this paper.

2. Financial crisis and the "root" of the problem

2.1 An outline of the financial crisis of 2008

Since the universal adoption of the capitalist system[1], severe global financial crises have been recurring in the last few decades. A common pattern can be identified in the spread of financial deregulation in most of the world markets that easily lead to the uncontrolled growth and subsequent bursting of speculative bubbles[2].
In the aftermath of every crisis, countries struggled to undertake emergency actions. They also tried to employ resources in order to prevent future crises in an attempt to learn from experience. In the short term, policy makers changed their priorities and committed to new and more virtuous strategies but the initial steps for reforming the system did not always convert into definitive commitments. That is why a serious discernment is pivotal in order to implement effective strategies in the long run.
Indeed, the "relative priorities of national policy making and international policies dialogue have been transformed by the 2007-08 crisis and its aftermath. The crisis expanded from the sub-prime loans market to the whole set of structured financial products, to the inter-bank market, to the global financial market. National and regional vulnerabilities became apparent, as the world experienced

---

[1] For more on this matter see, (B.J. Eichengreen 2008).

[2] "Oggi si conosce il periodo che va dall'anno 1998 al 2007 come quello della Grande Moderazione. Curiosamente coincide con l'acutizzarsi del fenomeno della "sregolazione" finanziaria. É il periodo della crescente confidenza nelle virtù dei mercati dove si crede che la concorrenza nell'ambito finanziario farà il suo compito insostituibile: raggiungere la autoregolazione degli stessi... L'idea dominante era la fede nella capacità dei mercati di stabilire da soli, esercitando liberamente le loro funzioni, i limiti del processo. Questa convinzione nella capacità di autodisciplina –autoregolazione- del Signor Mercato, aveva nel Presidente stesso del FED, Alan Greespan, il suo principale difensore... La sua particolare lettura, tanto espressamente ottimista, gli valse un grigio posto nella storia. Nel 2007 scrisse: "*...la proporzione del debito delle case rispetto al relativo reddito è in continuo crescendo e, gli opinionisti (critici) si sfregano le mani ...Queste paure (timori) ignorano un fatto fondamentale della vita moderna: in una economia di mercato, l'indebitamento (aumento del debito) cammina pari passo (va all'unisono) col progresso*" (Greenspan, 2007, pág. 298). Inoltre: "*L'aumento della leva finanziaria, sembra essere il risultato delle enormi migliorie nella tecnologia e nelle infrastrutture, non di umani notevolmente (significativamente) più predisposti al rischio*" (*Ibidem* pag. 415)" (De la Iglesia Viguiristi 2016).

further round of localized crises; this phenomenon was particularly evident in the case of sovereign debt crises in some European countries, with Union-wide institutional consequences" (Beretta, Zoboli 2012). Despite all the efforts of the bailouts of big credit institutions and emergency monetary and fiscal policies, the financial crisis soon became an unprecedented global recession.

Immediately after the Leman Bros failure, the existing global institutions revealed to be an "outdated muddle" (The Economist 2008) and their policies and rules completely ineffective. The first lesson learnt was that although financial markets are very important for the creation of wealth, being privileged intermediaries between lenders and borrowers, the "Efficient Market Hypothesis" (Fama 1965) is no longer a dogma[3]. According to this theory, assets' prices generated in financial markets always reflect their fair value. Nevertheless, this assumption is hardly consistent with the experience of speculative bubbles. As a consequence, policy makers should combine macroeconomic policies with market-regulation measures.

This is what happened with the adoption of the international agreements known as "Basel III" (2010), a comprehensive set of reform measures, developed by the Basel Committee on Banking Supervision, to strengthen the regulation, supervision and risk management of the banking sector[4]. Unfortunately, they are not binding agreements but only recommendations that need to be implemented by national systems.

At any rate, reforming a financial system is not only a financial problem however. This global crisis could represent an opportunity to better understand the transformation of markets and societies due to globalization and therefore the global dimension of economic governance.

---

[3] Samuelson held first that the stock market is "micro efficient' but not 'macro efficient". (Read 2013) After the crisis the EMH hypothesis has been strongly criticised for instance by former Federal Reserve chairman Paul Volcker: "it is clear that among the causes of the recent financial crisis was an unjustified faith in rational expectations and market efficiencies" (Volcker 2011).

[4] These measures aim to: improve the banking sector's ability to absorb shocks arising from financial and economic stress, whatever the source; improve risk management and governance; strengthen banks' transparency and disclosures. See www.bis.org

Intervening and regulating the extremely complex contemporary financial system is a titanic task. "Today, when global financial actors deal with global instruments in a variety of local situation and transnational policy instruments are missing, the nearest substitute available for global financial governance remains international cooperation, mostly consisting in supporting the hardest-hit countries and in adapting the «rules of the game» to the new crisis situation. The actual inability to prevent crises shows not only lack of motivations to cooperate and unwillingness to contribute, but more fundamentally a lack of «technology» for co-operating". "Financial systems are not «anonymous mechanisms», where risk aversion can be satisfied by just shortening the time horizon of financial connections. Back to basics, finance is an essentially relational phenomenon, and trust has to be gained by investing in meaningful personalized and durable relations. How this ultra-micro dimension connects to macro financial developments largely remains to be explored, but remains essential for a meaningful financial system to emerge" (Beretta, Zoboli 2012).

2.2. Brief critiques of the mainstream economic model

The financial crisis of 2008 has led to a profound and interrogative discussion on the sustainability of the capitalist system, specifically regarding criticisms of the fundamental principles of neoclassical economic theories which started in the second half of the nineteenth century and have lasted all throughout the twentieth, and have ultimately resulted in the so-called "mainstream" market economy. The market economy was based, initially, on freedom to choose and initiate business, private property, individual interest and the reasoning of human actions, the division of work, and perfect competition. Utilitarianism remains the framework of reference for the mainstream market economy, which postulates the single individual as perfectly rational and able to maximize their utility function (*homo oeconomicus*)[5]. The wellbeing of soci-

[5]. "Marginalism" (or the marginal revolution) characterizes the paradigm of the neoclassical economy in the face of classical economics. Its origins are usually traced from the works of three authors of the second half of the nineteenth century: Englishman W.S. Jevons with his *Theory of Poltical Economy* (1871); Austrian C. Menger with the

ety, therefore, "inadvertently" derives from the free trade between economic agents operating in order to maximize their own self-interest, resulting in a the collective and social balance that does not represent the individual rationality: the "invisible hand" of the market makes "efficient" exchanges in order to achieve the best possible result.
Therefore, a corollary of this approach cannot be a perfect anonymity of the internal interactions (Wicksteed 1910)[6] of the "catallactic companies" market (Pellicani 2017), namely, ensuring that each operator, in any given situation, (under income and budget constraints, alternatives, etc., in relation to the market price) would act in the same manner as any other in his place (almost like a "machine man" or "robot"). The outcome of such a mechanism can be appreciated in terms of "retributive justice," (exchange of equivalents) but it neglects the issues of "distributive justice" in society.

Principles of Economics (1871); and Frenchman L. Walras with *Elements of Pure Economics* (1874-1877). However, its international spread is mainly attributed to A. Marshall's *Principles of Economics* (1890), which became the dominant paradigm of the twentieth century amongst the "orthodox" economists as it featured microeconomic developments by Sraffa, Chamberlin, and Robinson, and that of the general equilibrium theory from Pareto, Hayek, Hicks, Arrow, and Debreu. At a macro level, it also speaks of "neoclassical synthesis" for which Marshall is an advocate, amongst several others such as Samuelson, and references to the principles of Keynesian economics. Classical economy differentiates itself from the neoclassical economy in its problems related to the subjective condition of economic choices and the theory of value: the first is based on "categorical" behavior, or social classes, and incorporates a theory of value that corresponds with "production costs" of the goods or services, (hence the Ricardian theory of value / labor and Marxist developments of surplus), and with the relative distribution of the value assumed by the product categories that have allowed for the realization of the human person (rent to the landowner, salary of the employee, the profit to the capitalist entrepreneur / investor, etc). The second is based on the individual methodology for which the decision-making process results in the economic choices of single individuals, taken as equally endowed with the same rationality coinciding with the end of the maximization of self-interest, and on a theory of value based on the "usefulness" of the goods or services for the consumer/user, and more precisely on the marginal utility that is able to make the last unit of the goods/services or fruits of labor consumed. Hence the notion of Marginalism." (Fortunato, 2013).

[6] The anonymity thesis belongs to Wicksteed, P.H., 1910.

The model of development, represented by the market economy, soon became subject to criticism and corrections.
Traditional criticisms have pointed out that an efficient market (which, as has already been mentioned, does not necessarily produce distributive equity) presupposes the realization of ideal conditions unlikely to be encountered in actual operations. In particular, it implies: a) perfect competition (full and effective) with free access to the single market; b) informative transparency; c) the absence of externalities, such as costs that are transferred on a third party (environmental costs, social costs, etc.). This ideal model, however, is not carried out in its purest form in daily practice: it instead requires continuous corrections, and undergoes "failures," especially in the production of "public goods," in part because of the "free rider" phenomenon in which the market economy does not allow it to operate.
At the communal level, (macro-economics) the principle of "capitalist accumulation," or continuous/unlimited growth, is not only in itself erroneous and unsustainable, but it proves to be destructive in the long run. It collides with the biophysical limit of natural resources and its related environmental costs, as well as with social limits because it causes dangerous inequality and dissolves social bonds. (Fortunato, 2013)
The most recent criticisms, based on sociological analyses, cognitive and experimental psychology, have questioned the anthropological foundation of the market economy. They note that (i) the motives of human action are more complex and less "rational" than the proposed ideal model, and (ii) the effective decision-making processes of man do not correspond to the abstract model of the *homo economicus*, the result of an extreme exercise of scientific reductionism.
Moreover, in the nineteenth century, the civilian economy based its market economy on an alternative and broader anthropological concept to the Smithian/Capitalist approach, known as the *homo reciprocans*. This model of economic behavior takes into account motivational heterogeneity characterized by altruism, fairness, solidarity, and reciprocity, or opposed negative feelings, but in any case oriented towards others (Eco, Fedriga 2015). Later, in the

mid-fifties, Herbert A. Simon (1947,1950, 1957) began to apply typical economic and political knowledge of what was then known as "behavioral psychology" (Watson 1913), which would lead him later to develop and elaborate the concept of "bounded rationality" (Simon 1982). "The decision of men are influenced by the objective limits of knowledge, an emotional component, the inability to foresee all the consequences, and incapacity to simultaneously consider all possible alternatives when there are too many variables at hand"[7].
In the wake of these "broader anthropological assumptions" of the second half of the twentieth century, three strains of studies were developed that proposed alternative scientific theories to the classic model of the market economy. a) The economics of happiness, which began in the mid-seventies of the last century from the psychological and sociological insights of recognized members of the scientific community. b) The theory of decline (happy or friendly), which includes amongst its most important philosophers, Serge Latouche. c) The economy of the gift, inspired by the famous *Saggio sul dono,* written by the sociologist and anthropologist Marcel Mauss between 1923 and 1925, found in Alain Caillé and the Anti-Utilitarian Movement in the Social Sciences, which started with its magazine in 1981. The most significant expressions of this theory, however, were strongly evoked in the encyclical of Benedict XVI Caritas in Veritate. (Fortunato, 2013)

3. New approaches to economics – from profit to value

The considerations regarding the overcoming of restricted and partial anthropological categories, not capable of enhancing the aspects of the intrinsic and distinctive relational nature of the human being as a totality, also have an impact on the choices and values they generate.
It is evident, therefore, that the maximization of economic value, pure and devoid of related elements, is no longer a sufficient goal.

[7] About the limits of the *homo oeconomicus* rationality of the neoclassic theory, see further Daniel Kahneman e Vernon Smith, Nobel in Economics 2002.

In order to be incorporated in the model of the choice of "other" items, one must choose a creation of value that is not limited only to economic and quantitative aspects. The creation of value, instead, should extend to create a broader social value that is not used purely as an economic driver.

The concept of business responsibility began to spread in response to the typical paradigm of free market societies. The "invisible hand" by Smith is no longer convincing because "although the marketplace has done a reasonably good job in deciding what goods and services should be produced, it has not fared as well in ensuring that business always acts fairly and ethically" (Carroll, Buchholtz 2012: 29).

Later, due to increasing business laws and regulations, emerged the *legal* model. Over time, society's expectations of business turned from being strictly economic to encompass also those other issues that previously were at the business's discretion: the model evolved into a *social* model, enhancing societal groups, a move from shareholders to stakeholders. It was the advent of Corporate Social Responsibility (CSR).

Murphy (1978) classified four CSR eras: the "philanthropic" era (period up to the 1950s) in which companies donated to charities; the "awareness" era (period 1953-67) in which it glimpsed the recognition of overall responsibility of business and its involvement in society affairs; the "issue" era (period 1968-73) in which companies started to focus on specific themes; the "responsiveness" era (from 1974) in which companies started to take oriented actions to address CSR.

Starting from the 1980s the evolution of the concept led to intend CSR as a base-point for developing complementary themes related to the relationship between business and society, and their mutual influences. By the 2000s the increasing emphasis on this theme in terms of theoretical contributions began to turn into empirical research on the topic and CSR definitely gained global expansion.

Early views of CSR state: "Corporate social responsibility is seriously considering the impact of the company's actions on society" (Paluszek 1976); "social responsibility [...] requires the individual to consider his [or her] acts in terms of a whole social system, and holds him [or her] responsible for the effects of his [or her] acts anywhere in that system" (Davis 1967).

Over time, the concept of corporate social responsibility gained considerable acceptance and broadening of meaning. The focus shifted from a general to a particular approach, privileging an orientation on specific issues.

The theme is very broad, nowadays a universally acceptable definition of CSR does not exist, and a study published in 2008 found 37 definitions of CSR (Dahlsrud 2008).

Despite these aspects, the distinguishing feature of the literature on CSR is that "it can be viewed as applying the stakeholder concept to non-traditional stakeholder groups who are usually thought of as having adversarial relationships with the firm. In particular, less emphasis is put on satisfying owners and comparatively more emphasis is put on the public or the community or the employees" (Freeman 1984)[8].

Among the different perspectives, it is relevant to mention that "the idea of social responsibility supposes that the corporation has not only economic and legal obligations, but also certain responsibilities to society which extend beyond these obligations" (McGuire 1963).

This interpretation is the basis to outline an extended framework identifying the four kinds of responsibility that constitute the CSR: economic, legal, ethical and philanthropic (Carroll 1979).

[8] In order to give a general overview of the focal contributes, some definitions of the concept are presented without demanding completeness. The active aspects of CSR in terms of protecting and improving are underlined by defining the concept as "the obligation of decision makers to take actions which protect and improve the welfare of society as a whole along with their own interests" (Davis, Blomstrom 1975). The World Business Council of Sustainable Development and the European Commission suggested formal definitions of CSR. According to the first entity, CSR is "the continuing commitment by business to behave ethically and contribute to economic development while improving the quality of life of the workforce and their families as well as of the local community and society at large" (WBCSD, 1998). According to the second entity, CSR is "a concept whereby companies integrate social and environmental concerns in their business operations and in their interaction with their stakeholders on a voluntary basis" (Commission of the European Communities, 2001), later intended as "the responsibility of enterprises for their impacts on society" (European Commission, 2011)

These aspects can be depicted through a four-layer pyramid, known as the "pyramid of CSR" (Carroll 1991), intended to illustrate that "the total CSR of business comprises distinct components that, taken together, constitute the whole" (*ibidem*, p. 42). Despite the usefulness of the separation into four levels in order to identify the different kinds of obligations, it is appropriate to underline that this view does not require a simultaneous fulfilment, since the components move "in a constant but dynamic tension with one another" (*ibidem*).

CSR is strictly linked to value creation, and in particular, both to economic and social value.

Economic value is precisely measurable, with different meanings according to selected dimensions. Conversely, it is not possible to determine social value from a quantitative point of view in a univocal way. Companies may adopt different indicators and ratings of social value (such as SROI-Sustainable Return of Investment, Dow Jones Sustainability Index, ASSET4 ESG, FTSE4Good), but none of these accurately measure corporate social performance and impact in a decisive way. This happens because ratings are not always consistent due to significant differences of criteria composing the indicators, and their different weight.

This consideration does not mean that good rankings are irrelevant, but emphasizes that quantitative indicators do not totally explain the impact of social value. To understand the deep meaning of social value, qualitative observations are needed.

In fact, value creation is a two-way process, encompassing both the value created by the firm to society and the value created by society to the firm. In other words, value creates value.

However, traditional approaches of both stakeholder and CSR theories fail to embed social/environmental considerations in the value creation process of a firm. The strategic approach to CSR, lately evolved into shared value (Porter, Kramer 2006, 2011), attempts to solve some issues related to the social responsibility of a firm and to overcome the questions associated to value identification and measurement, as emerged in the classical stakeholder theory (Crane et. al 2014).

Porter and Kramer (2006) argue that good citizenship, generic philanthropic contributions to local charities as well as mitigation of harms produced don't maximize the capacity of a company to generate value for society. On the contrary, a business should "identify the particular set of societal problems that it is best equipped to help resolve and from which it can gain the greatest competitive benefit".

Notably, by making social impact integral to the overall strategy, businesses embed the social value into their profit generation objective, according to a new capitalist paradigm, described as "profits that create societal benefits" (Porter, Kramer 2011). Their strategic approach to CSR, as opposed to a mere responsive approach, evolved into the famous concept of shared value, described as "policies and operating practices that enhance the competitiveness of a company while simultaneously advancing the economic and social conditions in the communities in which it operates". A company can therefore generate a shared value when it creates economic value by creating societal value. The shared value approach combines economic and social value, overcoming the research of a trade–off.

However, if this new approach can be reached only through a "reverse strategy" or social innovation, as written before, it could be possible that it requires the creation of a new company, as a spin off, or the investment into promising start-ups (following a corporate venturing approach). In fact, relevant or disruptive innovation is often impossible within the boundaries of established organizations (Vecchi, Cusumano, Brucconi 2014: 14).

4. New forms of finance: the impact investing

The anthropological typology that is both revealed at the economic level and will dictate the rules of the movement of money in decades to come is the financial system of the millennial generation, the largest interconnected generation the world has ever seen[9]. Over the next 30 to 40 years, $30 trillion in financial and non-fi-

[9] Millennials are the generation of the new millennium (also called generation Y) and they are those who were born between 1980 and 2000. They have the characteristic of being the first generation to have been born in the world of global communication where everything is connected and related.

nancial assets is expected to pass from the Baby Boomers to their heirs in North America (Accenture).
Hence, it is important to ask: What are the expectations of Millennials? What are the reasons for their economic action? What are the investment objectives of this new generation?
Having grown up in the age of connectivity, Millennials want to make a difference in the world and are eager to contribute positively to society and the environment. Witnessing the global recession in 2008 led many to be frustrated with the financial system solely focused on profit, and as they now inherit family fortunes or create their own wealth through entrepreneurship, they are demanding more options for impact investment (Monitor 2016).
The new generation is no longer interested solely in forms of philanthropy or classical alternative finance[10] but talks about broader financial concepts: impact investing.
The incentives that move Millennials towards impact investing are not the object of this paper. However, we take note and share the objectives of this preference and tendency in the new generations' choices of investments.
Impact investing is the current trend in the capital market. It is considered, on one side, as a new paradigm to cope with economic crises, curtailed public budgets and the answer to the more and more diversified needs of its citizens; and on the other side, it responds to private investors who are searching for new investment opportunities to channel the enormous liquidate available (Vecchi, Casalini, Balbo, Caselli, 2015)[11].
Impact investing may be the result of the drive to improve the performance of philanthropy (Porter, Kramer 1999) and the ascent of the so-called "philanthrocapitalism" (Bishop 2013; McGoey 2014; The Economist 2006) that brings 'hard-nosed' strategies,

[10] In this case, we refer to Ethical Banks or ethical reflection on finance. For more details see (Sangalli S. 2014: 261-267).

[11] "Globally, private wealth has never been so high: in 2013 total global financial assets grew to $225 trillion, tripling the world's GDP (McKinsey Global Institute 2014), even if only 22%of them is represented by equity investments, whose CAGR in the period 2007 – 2012 was -5,5%; high net worth individuals (hereafter HNWIs) financial wealth reached the peak of $ 52.6 trillion worldwide in 2014, of which 13,5% is invested in alternative".

performance metrics, innovative financing models that aim to 1) make philanthropy more effective; and 2) make it a more lucrative industry in itself (McGoey 2014). Specifically, impact investing seems to pursue shared value, here it is understood as a value that incorporates economic value and social value at the same time.
The term "impact investing" was coined in 2007 by Rockefeller Foundation that defines it as "a set of investments in society, organizations, and funds with the intention of generating, in addition to financial returns, a quantifiable impact at both a social and environmental level". The term "intention," in the present context, differentiates such investments from those that are merely socially responsible. These last investments look to avoid social and environmental damage derived from resources that otherwise would be directed only to one scope: profit. A social impact investment can be made in either developed or developing countries, with the objective of procuring financial returns either below or in line with market levels. This characteristic trait distinguishes impact investing from ordinary contributions or donations that do not expect any financial return (Martin 2014).
Although introduced by the Rockefeller foundation back in 2007 as a "worldwide industry for investing for social and environmental impact," the definition of Impact Investing has changed since it was first proposed ten years ago. Numerous institutions, scholars and academies[12] have provided their own definition of the Impact

[12] Many academies have launched research programs and study courses. In England both Cambridge and Oxford take impact investing into serious consideration with the Oxford Said Business School also offering a program dedicated to it. On the other hand, Cambridge University with the Institute for Sustainable Leaders has chosen a more academic way to face this subject by producing studies and reports. A similar activity is carried on at Stanford University where the Stanford Social Innovation review reunites all the important papers and researches on this topic and giving on either theoretical aspects as well as more concrete ones such as its evaluation in balance sheets.
Opposite to this, and more similar to the Oxford Said one, are the projects of both Columbia and NYU Stern University which provide programs and classes to strongly preparate today and tomorrow managers to face this important challenge. In Europa, Bocconi University with its impact investing lab is not to be considered lower to UK/US universities. Bocconi's lab main characteristic is not to provide an academic, abstract impact investing approach but to cooperate with companies and entrepreneurs to define its activity perimeter and to elaborate an counting method to better understand its proper

Investing to the business world, but in doing so, they have only added more complexity to the concept. The latest definition of the concept, once again provided by the Rockefeller Foundation in 2014, has been defined as "A middle way between philanthropy and pure financial investment. A means of using capital to drive financial value and social environmental impact simultaneously." Confusion about Impact Investing exists not so much in the understanding of its definition, but more so in how it is different in its nature, distinguished features, and outcomes from the methods used in the past. The simple answer is that Impact Investing is meant to be more sustainable than its predecessors. Guidelines distinguishing the features of Impact Investing have been set by entrepreneurs Brest and Born in 2013, in order to ensure that impact investments both help to better improve the common good, and further differentiate this concept from additional forms of social investments:

1. Intending social and/or environmental impact, making it different from a purly financial investment.
2. Additional, which means that the investments must increase the quality or quantity of social or environmental outcomes beyond what would otherwise have occurred in the case of a traditional investment;
3. Implies the generation of financial returns that marks its difference with the philanthropic approach (Vecchi, Cusumano, Brucconi 2014: 5).

Impact investing is also ranking high in the policy agenda of governments and international organizations. In 2013, the UK arranged within G8 a dedicated taskforce that suggests areas, criteria and tools for impact investing in Europe[13]. In addition, the World Economic Forum launched an initiative to shed light on this new phenome-

value creation capacity.

[13] For further details see "Impact investment: the invisible heart of markets", Report of the Social Impact Investment Task Force established under the UK's presidency of the G8, 15 September 2014 and relating "La finanza che include: gli investimenti ad impatto sociale per una nuova economia", Rapporto Italiano della Social Impact Investment Task Force.

non. The European Union, through the European Investment Fund (EIF), the European Investment Bank Group's specialist risk capital arm, launched the Social Impact Accelerator (SIA) initiative to foster the development of a European market for impact investing. Also, global financial institutions, like J.P. Morgan and Credit Suisse have been the main promoters of this new investment approach. In 2008, in the wake of the financial crisis, these organizations alongside the Rockfeller Foundation launched the Global Impact Investing Network (GIIN), which almost 200 members.

4.1. Areas and Players.

Impact investing was born to answer the demands of both the market and society. It takes its place in those areas where, due to high risks, neither the State, nor welfare, nor for-profits invest, that is where the need of services to individuals are not adequately sustained by the support of public entities. This provides investment solutions for the four megatrends that are transforming businesses, governments and civil society by creating value and financing the common good:

- A consistent and unexpressed demand at the "Bottom of the Pyramid," is made up of 4 billion people, whose annual income, measured in local purchasing power, is below $3,000 USD. Such a mass of consumers and producers potentially moves an economy that already represents $5 billion USD. Therefore, the funding provided, in part by traditional market capital and in part by public/private partnerships, helps these subjects to enter the market, thereby achieving satisfactory living conditions and stimulating national economies.
- The development of a green economy is necessary to reduce environmental risks and ecological issues, and thus human wellbeing and social equity.
- The restructuring of the welfare state to increase efficiency from the moment that part of public spending is greater than revenue: this requires the intervention of other forms of financing to address the most pressing social challenges and to free the State from the corresponding expenses;

- Consolidating the LOHAS "Lifestyles of Health and Sustainability" segment; these consumers are willing to spend more on products that reflect their own values (environmental, sustainable and socially responsible) (Martin, 2014: 13).

Among the areas of impact investing's intervention, we find microfinance, agriculture, nutrition, health, education, justice, employment, social housing, energy conservation, and financial services.
Despite being at the experimental stage characterized by fragmentation and being undersized, along with the presence of many "small" innovative players entering market niches, impact investing has excellent potential that can be expressed only if investors and intermediaries will act in a harmonic manner to create an ecosystem capable of aggregating supply and demand from various sources.
The main actors of the system whom are called upon to take up this challenge are the philanthropic foundations, which with the first investment programs experienced in the US in the 60s financed with donations companies and business that shared their social mission. Philanthropic foundations have laid the bases for what today is the phenomenon of impact investing, and continue to be the "healthy carriers" of its best practices.
Other key players in the impact system are "angel investors" and "early stage" investors. These are able to support companies in the early stages of their development (seed, start-up and expansion) and hence can bridge the gap between funding during the launch phase by the founders and the venture funds of later stages of business development.
Another category of players in the impact investment market is that of professional investors, which includes high-net worth individuals (HNWIs - individuals whose total personal income and assets, excluding their residences, exceed one million dollars) and family offices (organizations engaged in managing the assets of one or more families with high availability, and providing advice regarding investment management). These individuals have so far been the main source of funding for impact investing operations. Their distinctive characteristics compared to other capital provid-

ers are flexibility, a high level of prudence and independence in taking investment decisions.
We must also consider the category of institutional investors, i.e. operators acting as intermediaries for investments on behalf of third parties (Sim Fund, asset management companies, banks, pension funds, banking foundations). So far, all these parties have been less active in terms of impact investment because they are less inclined to assume risk and therefore they are wary given the lack track records for this type of investment. Another source of capital for the impact investing sector comes from corporate impact venturing. This is the process by which for-profit organizations can adopt new philanthropic approaches as developers of social enterprises through spin off operations, providing expertise, resources and technologies, so that they can remain on the market and generate social impact.
Intermediaries, whose presence is crucial in all financial markets, hold a fundamental role to bridge supply and demand in social impact investments. In traditional finance, intermediaries act as links in transactions and include investment banks, brokers, and other players, often of large dimensions. However, in the field of impact investing smaller and highly specialized subjects tend to prevail, although there has been the entry of large investment banking. Some of these include JP Morgan Social Finance and Deutsche Bank, and the establishment of special platforms (e.g. "investing with impact" by Morgan Stanley and "ImpactBase", managed by the GIIN) which provide information and aggregate data for investors including the possibility to see the social impact of investments by sector, region or asset class.
The importance of intermediaries in the ecosystem of impact investment has been well understood in the UK, which in 2012 set up Big Society Capital (BSC), an investment company that operates as a wholesaler in the English capital impact investment market. It has assets of about 600 million pounds made up in part from dormant bank deposits and other funding from major British banks (Barclays, HSBC, RBS, Lloyds). It invests in financial intermediaries that will help social enterprises to raise capital and investors to find good investment opportunities.

From the above we can conclude that impact investing is a market in development and is still not an ecosystem with clearly defined boundaries. The methods of investment can range from direct investment in social enterprise loans to brokers and forms of capital spanning from venture capital and debt capital (predominantly) and mezzanine debt or a combination of these instruments, the so-called hybrid capital. With structured finance instruments, while it is possible to create a whole range of financial products, it will be however state's participation with the availability of public funds to take the lion's share.

4.2. Some tools for Impact Investing

As part of the innovative finance, for several years, some tools were created to aggregate private funds aimed at achieving positive social impacts. In the institutional context, one can highlight the social impact bond (SIBs).

SIBs are multi-stakeholder arrangements between a public authority, investors, a service provider and an intermediary organization. A bond is sold to private investors who are paid a return only if the associated social project succeeds in attaining a certain level of social result within a given period of time. More specifically, capital provided by private investors is used by one or more service providers to set up or finance activities that will both bring about positive social outcomes for the beneficiaries and reduce future public costs. These cost savings are then shared by public authorities (sometimes referred to as the commissioner of the bond) with the original investors, who receive payment in the form of a return on their original capital investment. Usually an intermediary or advisor brings together these three players and helps to set up the necessary structures in return for a fee. In principle, both the investors and the public benefit from this 'win-win situation'. Best of all, no public funds need to be raised or committed during the set-up and operation of the service, i.e. until the desired results are achieved.

SIBs are a form of 'pay for results' funding because the repayment of the original investment and/or payment of financial return is

dependent on the achievement of measurable social outcomes. If the service attains a certain goal (i.e. social outcomes improve), then at the end of the term of the bond, the original investment is repaid to the investors plus a financial return. This return may vary depending on the extent of the improved outcomes, with higher returns paid for better results. As is evident from this description, the name 'bond' is in fact a misnomer: social impact bonds are not like traditional bonds, i.e. debt securities that pay out a fixed rate of interest until maturity. They are more like equity investments that pay out only on the basis of results achieved, and they carry high levels of risk: social investment bond-holders may lose their entire investment if the project is not successful (Davies 2014).
The first SIBs were issued in 2010 by Social Finance Ltd in the UK for the social program of Peterborough Prison. The bond raised 5 million pounds from 17 social investors to fund a pilot project with the objective of reducing re-offending rates of short-term prisoners. The relapse or re-conviction rates of prisoners released from Peterborough was compared with the relapse rates of a control group of prisoners over six years. If Peterborough's re-conviction rates proved to be at least 7.5% below the rates of the control group, investors receive an increasing return that is directly proportional to the difference in relapse rates between the two groups and is capped at 13% annually over an eight-year period.
In the wake of the results achieved by the first trials in the UK, especially the SIB of the Peterborough prison, some administrations in the US have decided to launch similar pilot projects. We refer in particular to the SIB of New York City. This was started to reduce the recurrence rate of young detainees, and the two PFS programs in the State of Massachusetts, each launched in partnership with Third Sector Capital Partners and addressed to homeless and young people with the goal of reducing the crime rate. Also in Australia, the State of New South Wales has promoted in November 2010 a feasibility study in partnership with the Center for Social Impact - a research group involving three different universities of the country - and the Young Foundation for the experimental release of SIBs. This process has led to an issuance of $ 25 million in March of 2011, divided into three tranches for prevention ini-

tiatives in disadvantaged areas with a high juvenile crime rate, and to support households at risk of poverty and social exclusion and, ultimately, for services to people with disabilities.

In Continental Europe, the SIB does not seem to have yet convinced governments, intermediaries and investors, although the interest and political commitment to the theme of social entrepreneurship and its funding has proved, in both intentions and deeds, to be considerable. In particular, the European Commission has promoted a number of significant initiatives. The first refers to the Social Business Initiative, by which it decided to place at the top of its agenda the theme of social entrepreneurship as an effective tool to combat uneasiness and social exclusion and to promote competitiveness, at an international level of the European model of socially inclusive market economy (Fondazione Cariplo 2013).

4.2.1. Italian Tools

At the business level, social entrepreneurship is the target of impact investors. The combination of financial sustainability and social impact are often merged into social enterprises (Austin, Stevenson, and Wei-Skillern 2006; Dees 1998; Peredo and McLean 2006; Wallace 1999; Weerawardena and Mort 2006) or can be made possible through social innovation (Kanter, 1999; Mulgan, et al., 2007). According to Dees (1998) a social enterprise is mission-related, and wealth creation is just a means to reach the desired impact. The European Commission (2011) posited that social enterprise is an operator in social economy, whose main objective is to have a social impact rather than make a profit for their owners or stakeholders. These definitions do not exclude the possibility of profit generation, as ventures created by social entrepreneurs can certainly generate income, and they can be organized as either not-for-profits or for-profits (Martin and Osberg 2007). Behind different definitions, the social enterprise's concept captures and describes organizations that adopt managerial models and tools to better answer societal needs. Therefore, the suggestion (Nicholls 2006) to approach social enterprises through the entrepreneurial and management lenses in order to overcome academ-

ic or judicial speculations may be useful, though it is important to understand the peculiar differences with traditional businesses (Vecchi, Cusumano, Brucconi 2014).
The Italian legal system has been improved by the 2016 Stability Law (Law no. 208, December 28th 2015) with a new corporate structure that put a new feature at the side of the 'traditional' non-profit: the "Benefit Company".
The legislator meant to highlight the social purpose carried out by this new form of business, as showed by the six paragraphs entirely related to the latter, especially where it stated that "without prejudice to the lucrative feature typical of the profit-sharing" this company shall be focused on "one or more common benefits".
The purpose of a "common benefit" is not so innovative in itself, as it is typical of all that pertains to the third sector, such as the social enterprise provided by the Legislative Decree no. 155/2006.
The peculiarity of the benefit company may be understood through the dichotomy of the scope: if it was aimed just to create a public edge, without any form of profit sharing, the legislative intent would not be satisfied at all.
The regulation, therefore, sets the way through which a company shall behave towards several players, being provided that it happens "in a responsible, sustainable and clear way towards people, community, landscape, environment, cultural and social goods and services, bodies, associations and other stakeholders".
This new legal subject is deemed to be unfamiliar with current regulation: in fact, even if the Stability Law does refer to the corporate models under Book V of the Italian Civil Code, the balance between profit and no-profit is likely to put distance between the provisions of both Article 2247 of the Italian Civil Code and the Legislative Decree no. 155/2016.
In order to better analyze this corporate form in its essential features, it has to be said that legal status is one of the traditional corporate forms, being different from the latter in having a double case, as a sort of *negotium mixtum.*
Delving deeper, one must mention that the analysis carried out by some of the best authors, is based on the distinction between two main features: first, the subjective profit, pointed towards

the shareholders; second, the objective profit, directed at the company.
The idea under the benefit company is to change the concept of profit, which is the soul of corporate issues, spreading it over third parties, and maybe putting it outside economic or business activity. What precedes represents the real innovation, as it has been introduced as a third way between profit and no-profit: this is the new corporate form that shares not only profits but also benefits and utilities.
The particular attention granted by the Italian Legislator providing such an innovative legal tool deserves to be mentioned, especially in comparison of the previous western models. For several years, exactly by 2010, "benefit corporations" or "B-corporations" have been developed in several countries in the US, above all Maryland. These forefathers of the Benefit Company are aimed to create welfare, not only profit. Chairmen and directors are called to develop this purpose, otherwise being responsible under the benefit enforcement proceedings (BEP), that can be filed by any shareholder where one of these goals no is reached.
The main difference with our domestic benefit company lies in the criteria governing periodical assessments of the purpose's achievement: the systems both externalize this activity, but in the US the criteria are elected by the same company, whereas in Italy they are provided by Law.
In addition to the innovations introduced by the legal system, as for those active in the investment market, in Italy was born one of the first investment funds devoted entirely to impact investing: Oltre Venture. The first Italian Impact investment Fund was founded in 2006 with the aim of supporting the strategic development of "not for profit" organizations (Oltre Venture I).
Oltre Venture I represented a first attempt to overcome the traditional philanthropic approach towards the social sector and has demonstrated that impact investing could be an opportunity to channel new and sustainable businesses and the enormous liquidity available in Europe and worldwide. The investment's targets were the following: real estate investments for social activities (social housing) and services (healthcare, microcredit, social care)

with a mix of greenfield (seed/start-up phase) and brownfield/expansion (investments into already existing social businesses). The fund target size was € 12 million, with a duration of 10 years and an investment period no longer than 4 years.

Oltre Venture I reached a total commitment of € 7.5 million from 22 equity investors, mostly HNWIs, entrepreneurs and foundations, who accepted this challenge mainly for philanthropic purposes. Due to the small size of the fund, the operational expense was covered by the founder and not by a management fee, as it generally happens in traditional funds.

Oltre Venture I focused on 3 main investments, representing 66% of its total portfolio: PerMicro spa (microfinance), Sharing srl (social housing) and Società e Salute Srl (Healthcare) (Vecchi, Casalini, Balbo, Caselli, 2015: 14)[14].

After the positive results achieved by the Oltre Venture Fund I, Oltre Venture Fund II was created with a target size between € 20 to 25 million, of which € 10 million was invested by the European Investment Fund. According to the target risk-adjusted IRR negotiated with EIF, Oltre Venture II will invest:

- where public services can't meet the evolution of population needs, such as in the healthcare, education, professional training, employment, and student houses;
- where there is a potential of innovation and an expected growing demand, such as in agriculture and food distribution, tourism, lighter and more flexible residential care for the elderly;
- where the lack of capital hampers economic and entrepreneurial development, such as in southern Italy, and suburbs of large cities.

[14] Oltre Venture I portfolio combined investments with different profiles. For example, PerMicro can be regarded as a hybrid investment, since BNL bank supports PerMicro through its CSR budget. For this reason, this investment is barely replicable and it can be considered a typical venture philanthropy investment. On the contrary, Società e Salute is a financially free standing business, fully replicable, which has actually represented the reference case for the fund raising of Oltre Venture Fund II, where the EIF has invested € 10 million.

Oltre Venture II will certainly have lower risk compared to Oltre Venture I, which made pioneer investments. However, even despite lower risk, success may be challenging for Oltre II. Infact, public companies, big private corporates and banks generally prefers traditional CSR to Impact Investing, because if the CSR investment/project does not generate any relevant impact it can be however considered an effort to do good. On the contrary, if the Impact Investing does not reach the target IRR, despite the impact generated, it may be considered a wrong investment and therefore able to negatively affect the reputation of the manager who took the decision and of that of the company, especially when it is funded by the budget dedicated to CSR initiatives.
The Oltre Venture case clearly shows that the attraction of stable financial resources to sustain the development and expansion (or scalability) of organizations aimed at tackling some societal needs requires business models capable of reaching certain levels of financial returns, coherent with the risk embedded.
Beyond the precise characteristics and growth choices employed by Oltre Venture, this fund demonstrates that "it is possible" to combine profit and impact for society (Vecchi, Casalini, Balbo, Caselli 2015, 17:19).

4.3. How new technology contributes to the development of Impact Investing.

Despite the fact that impact investing has been welcomed by various institutions and financial realities, some resistance still exists towards its development, specifically from the bank system on the topic of the investment criteria aimed towards profit maximization. Between new technology and a changing network, cash-exchange circuits are being generated that are completely bypass the banks and financial institutions. The proliferation of peer-to-peer platforms, the phenomenon of crowdfunding, and the use of Bitcoin and the relative blockchain processes, although moving only small monetary sums, are an example of this.
Crowdfunding is the practice of funding a project or venture by raising capital from a large number of people, through the collec-

tive effort of friends, family, customers, and individual investors. This approach taps into the collective efforts of a large pool of individuals—primarily online via social media and crowdfunding platforms—and leverages their networks for greater reach and exposure. There are 3 primary types of crowdfunding: donation-based, rewards-based and equity crowdfunding and in 2015, it was estimated that worldwide over US$34 billion was raised this way (Forbes, Barnett 09/06/2015).

Crowdfunding has been used to fund a wide range of for-profit entrepreneurial ventures, in particular seed start-ups, such as artistic and creative projects, or community-oriented social entrepreneurship projects.

Bitcoin is a digital currency (also called crypto-currency) that is not backed by any country's central bank or government. Bitcoins can be traded for goods or services with vendors who accept Bitcoins as payment. It is the first decentralized peer-to-peer payment network that is powered by its users with no central authority or middlemen. Bitcoin was the first practical implementation and is currently the most prominent triple entry bookkeeping system in existence.

Bitcoin can be used to buy things electronically. In that sense, it's like conventional dollars, euros, or yen, which are also traded digitally. However, bitcoin's most important characteristic, and the thing that makes it different from conventional money, is that it is *decentralized* (no single institution controls the bitcoin network but the network of users) and allows trade without commissions or transaction fees.

The problems of transparency of the bitcoin system could be overcome by a new system called blockchain.

The blockchain is seen as the main technological innovation of Bitcoin, since it stands as proof for all the transactions on the network. A block is the 'current' part of a blockchain which records some or all of recent transactions, and once completed goes into the blockchain as a permanent database. Each time a block gets completed, a new block is generated. There is a countless number of such blocks in the blockchain. So are the blocks randomly placed in a blockchain? No, they are linked to each other (like a chain) in proper linear, chronological order with every block con-

taining a hash of the previous block.
To use conventional banking as an analogy, the blockchain is like a full history of banking transactions. Bitcoin transactions are entered chronologically into a blockchain just the way bank transactions are. Blocks, meanwhile, are like individual bank statements. In potential, these tools could facilitate the matching of supply and demand in the impact investing market without the use of formal financial intermediaries.

5. Impact investing in Vatican

Impact investing is no stranger at the Vatican.[15] In the course of the last three years, the Vatican has put on two series of conferences on the topic in order to better understand and apply this new way of investing. The first two-day impact investing symposium, "Investing for the Poor," 2014, was "an opportunity to learn core concepts of impact investing, to discuss how it aligns with Church mission, and to discern how the Church might use or promote impact investing to serve the poor."[16] The second Vatican conference on impact investing, subtitled, 'Making the year of mercy a year of impact for the poor,' "presented an opportunity for dialogue, learning and networking among impact investing thought leaders and Catholic leadership."[17]
As evidenced in the preceding paragraphs, one of the problems with impact investment is that since it is a new concept not everyone agrees on how it should be defined. In order to better understand the way Pope Francis sees this field, it is important to present the words he gave during his opening address to the first series of conferences. He stated the following:

[15] Here we must qualify this term. The 'Vatican' as such is a complex reality. Here, to be more specific, we are referring to two series of conferences put on by the, then called, Pontifical Council for Justice and Peace (PCJP), Catholic Relief Services, and the Mendoza College of Business, University of Notre Dame. The first of these conferences was preceded by an address of the Holy Father as quoted further below. We also clarify that this is one of many dicastries and impact investment is one of the many initiatives the 'Vatican' presents.
[16] See www.viiconference.org.
[17] See *ibidem*.

> *"Impact investors are those who are conscious of the existence of serious unjust situations, instances of profound social inequality and unacceptable conditions of poverty affecting communities and entire peoples. These investors turn to financial institutes which will use their resources to promote the economic and social development of these groups through investment funds aimed at satisfying basic needs associated with agriculture, access to water, adequate housing and reasonable prices, as well as with primary health care and educational services. Investments of this sort are meant to have positive social repercussions on local communities, such as the creation of jobs, access to energy, training and increased agricultural productivity. The financial return for investors tends to be more moderate than in other types of investment."*[18]

This last quote is important due to the Pope's famed suspicion towards market economies and capitalism. Here he seems to see a path that can help avoid the many pitfalls or market failures in relation to the poor and the environment that he has pointed out in the years of his pontificate. However, the very fact that these two symposiums have been organized under his auspices is quite indicative of his openness to this new way of market financing. He points out that, "the logic underlying these innovative forms of intervention is one which acknowledges the ultimate connection between profit and solidarity, the virtuous circle existing between profit and gift … Christians are called to rediscover, experience and proclaim to all this precious and primordial unity between profit and solidarity."
Beyond only discussing the nature and promise of impact investing there were also a few panels dedicated to its actual implementation in the Church today. During one of these panels[19] named, 'Investing Catholic Funds for Impact', the general treasurer of the Society of Jesus, Fr. Thomas McClain S.J., described himself as a fussy customer for impact investing. This is in part due to the novelty of this new way of investing and the difficulty to really making sure the companies that call themselves such are truly living up to this title. He explained how he has accompanied the process of moving from only negative screens in investing to impact investing, or as he calls it, positive screening. Fussy customer

[18] *Ibidem.*
[19] *Ibidem.*

or no he stated that 10% of the Jesuit international fund (worth around $200 million) is reserved for impact investing.
Another Catholic group that is actively involved in impact financing is the Missionary Oblates of Mary Immaculate. Fr. Séamus Finn represented this group as Chief of Faith Consistent Investing for the Oblate International Pastoral (OIP) Investment Trust. This Catholic fund, already including 16 negative screens, distinguishes two groups of positive investments within its impact investing. The first group, worth $50 million, is invested in Private Equity impact funds, while the second group, worth $4 million, is invested in concessional rate impact funds.
A third catholic organization run by Dale M. Hunt, the Managing Director of Private Equity for Ascension Investment Management (healthcare fund), shared that this fund has invested $43 million of its $50 million in impact investing as well. With these three examples, we can see that the Catholic Church has groups who are actively committed to impact financing.
However, it was surprising to see that the *Istituto per le Opere Di Religione* (IOR) or commonly known as the 'Vatican Bank,' whose president of the Board of Superintendents is Jean-Baptiste de Franssu admitted that it doesn't have a current impact financing portfolio. This is due in part to its recent restructuring, its history of not investing in the private sector and its being completely dedicated to services such as managing the economic activity of the Holy See and managing some congregation's and diocese's wealth. This being admitted, Mr. De Franssu mentioned that the IOR is presently considering how it could get involved in impact financing.
All of this being said we come to see a picture where the Church in some sectors is involved in impact investing directly, even its own 'bank' is considering how to begin, and one of its dicastries has put on two series of conferences to better understand and implement this new form of investing.

6. Conclusion

The Catholic Church began to specifically "show interest" in economics in 1891 with the encyclical letter "*Rerum Novarum*" in light of three events. These are: (i) the Industrial Revolution, (ii) the tremendous impact of Karl Marx's works that on the one hand affected the future development of classical economics and on the other hand injected impetus to widespread social discontent, (iii) the failure of the academic community to come up with an economic theory able to interpret the new events, providing an effective alternative to the requests coming from the labor markets (Pasinetti 2010).

Today the deep changes brought by the Technological Revolution at a global level lead us to reflect on the need to provide concrete and appropriate answers to our economic system that is again at a fundamental crossroads. We firmly believe that, contrary to the past, all institutions, including the academic world and the Vatican, must participate in the edification of a new economic-financial system that is already emerging in a realistic and altruistic manner.

Impact investing actually represents the most concrete attempt to fill the abovementioned gap. It seems to have those characteristics that can be summarized as equity, solidarity and reciprocity. The Social Doctrine of the Church has always suggested that the development of a sustainable and inclusive economy must be aimed at achieving the common good.

The Vatican, as well as Europe and Italy, have begun to take seriously this market trend. Drawing attention to the issue, it has promoted and recognized established practices.

At any rate, the development of impact investing will depend on the ability of all actors involved in the process to have a strong influence on the actual debate, getting their hands dirty and having the courage to go effectively "against the system".

References

BERETTA S. – ZOBOLI R., *Crisis and change – The Geopolitics of Global Governance*, Vita e Pensiero, Milano, 2012.

CARROLL A. B., "A Three-Dimensional Conceptual Model of Corporate Performance", *The Academy of Management Review*, Vol. 4, n. 4, 1979.

— «The Pyramid of Corporate Social Responsibility: Toward the Moral Management of Organizational Stakeholders» in: *Business Horizons* 34(4), 1991.

— with BUCHHOLTZ A. K., *Business and Society: Ethics, Sustainability, and Stakeholder Management*, CENGAGE Learning, Stamford, 2012.

COMMISSION OF THE EUROPEAN COMMUNITIES. *Promoting a European Framework for Corporate Social Responsibilities*, COM 2001, 366 final, Brussels.

DAHLSRUD A., «How Corporate Social Responsibility is Defined: an Analysis of 37 Definitions» in: *Corporate Social Responsibility and Environmental Management*, 15, 2008.

DAVIS K., "Understanding the Social Responsibility Puzzle, *Business Horizon* Winter, 45–50, 1967.

— with BLOMSTROM R. L., *Business and Society: Environment and Responsibility*, 3d ed., New York: McGraw-Hill, 39, 1975.

DE LA IGLESIA VIGUIRISTI F., *La crisi economica attuale: aspetti economici, etici e politici*, Quarta Sessione, Università Gregoriana. Pro-Manuscripto

ECO U.- FEDRIGA R., *La filosofia e le sue storie, L'età contemporanea*, Editori Laterza, 2015.

EICHENGREEN B.J., *Globalizing Capital. A History of the International Monetary System*, Princeton University Press, 2008.

EUROPEAN COMMISSION, *Communication From The Commission To The European Parliament, The Council, The Europe-*

an Economic And Social Committee And The Committee Of The Regions. A Renewed Eu Strategy 2011-14 For Corporate Social Responsibility, Brussels, 25.10.2011 COM (2011) 681 final.

EUROPEAN PARLIAMENTARY RESEARCH SERVICE (Author DAVIES R.), *Social impact bonds, Private finance that generates social returns,* Briefing August 2014, PE 538.223.

FAMA E., «The Behavior of Stock Market Prices» in: *Journal of Business* Vol. 38, No. 1, 1965 Jan.

FORBES, Barnett C., *Trends Show Crowdfunding To Surpass VC in: 2016*, 09/06/2015, www.forbes.com

FREEMAN R. E., *Strategic management: A Stakeholder Approach*, Marshfield, MA: Pitman Publishing, 1984

FONDAZIONE CARIPLO, «I Social Impact Bond, La finanza al servizio dell'innovazione sociale?» in: QUADERNI DELL'OSSERVATORIO n.11, 2013, www.fondazionecariplo.it/osservatorio.

FORTUNATO S., «Crisi, economia di mercato e modelli alternativi: provocazione per il giurista» in: *Orizzonti di diritto commerciale*, 2013.

MAXIMILIAN Martin, *Come rendere appettibile l'impact investing*, working paper ed. Impact Economy Vol. 4, Edizione Italiana ed. Fondazione Cariplo, 2014.

MCGUIRE J.W., *Business and Society*, 144, McGraw-Hill, New York, 1963.

MONITOR, «Narrative Analytics on Impact Investing, Global Impact Investing Network» in: *Monitor 360*, Omydiar Network, 2016, February.

MURPHY P. E., «An Evolution: Corporate Social Responsiveness», *University of Michigan Business Review*, 1978, Nov.

PASINETTI L., *Dottrina sociale della chiesa e teoria economica*, Relazione presentata nella Sessione di apertura del VII Simposio Internazionale dei Docenti Universitari organizzato dal Vaticano di Roma, 24 giugno 2010.

PELLICANI L., *La Genesi del Capitalismo e le origini della Modernità*, Rubbettino, Soveria Mannelli, 2006.

READ C., *The Efficient Market Hypothesists: Bachelier*, Samuelson, Fama, Ross, Tobin, and Shiller, Palgrave Macmillan, 2013.

SIMON H.A., *Administrative behavior; a study of decision-making processes in administrative organization*, Macmillan Co., New York, 1947.

SIMON H.A., *Public administration*, Knopf, New York, 1950.

SIMON H.A., *Models of man: social and rational; mathematical essays on rational human behavior in society setting*, Wiley, New York, 1957.

SIMON H.A., *Models of bounded rationality*, Cambridge, Massachusetts, MIT Press, 1982.

SMITH A., *The Theory of Moral Sentiments*, Part IV, 1759.

The ECONOMIST, *What a way to run the world*, 5/07/2008.

VECCHI V. – CASALINI F. – BALBO L. – CASELLI S., «Impact Investing: a new asset class or a societal refocus of Venture Capital?» in: *Position Paper, Impact Investing Lab*, in SDA Bocconi School of Management, 2015.

VECCHI V. – CUSUMANO N. – BRUCCONI M., «Impact Investing: Beyond CSR and Social Enterprise» in: *Position Paper, Impact Investing Lab* in SDA Bocconi School of Management, 2014.

VOLCKER P., «Financial Reform: Unfinished Business», New York Review of Books, 2011.

WATSON J.B., «Psychology as a behaviorist view», in *Psychological Review*, n.2/20, 1913.

WBCSD, (1998, Sept 6-8), *Stakeholder Dialogue on CSR*, The Netherlands, 1998.

WICKSTEED P.H., *The common sense of political economy, including a study of the human basis of economic law*, 1910, Macmillan and Company, limited, 1910.

Workshop n. 5

COSMOLOGY After Quantum Physics: Shaping a New Cosmology Case Study: Cosmic Inflation and Fine Tuning

Lucia Cosmano, Marilena Pisani,
Michael Baggot LC, Sara Bindi

with
Juzyszyn Joanna, Gioele Tornaghi,
Valentina Zavarella.

Under the supervision
of Dr. Umberto Cannella
Swiss Space Center,
Polythecnic Federal Institute (Lousanne)

1. Introduction

The stars, the planets, space, earth and its origins and end are topics that seem to have little to do with the technological revolution and the rapid changes of our society. They are subjects that we instinctively perceive as distant and immutable. One could ask what relevance the discovery of a spherical Earth or the arrangement of the solar system might have upon daily life or the international financial market. Nevertheless, man continues to lift his gaze toward the heavens and question the meaning of the universe, the Earth, and his own life.

When man pauses from his quotidian concerns to consider the cosmos, he is introduced into a profound and concrete discussion about the limits man has set upon himself for centuries, the essence of human nature, and the direction of our civilization. History has shown that the scientific revolutions wrought through the theories of Galileo, Copernicus, Newton, Einstein, and others have philosophical and cultural consequences that extend well beyond the confines of the professional scientific community. It would, however, be unfair to reduce the great cultural advances to the extraordinary intuition of a few men. Rather, the great minds impacted history not only through their individual gifts, but also in assembling and concertizing the labor of those who preceded them. As Isaac Newton himself acknowledged, "If I have seen further it is by standing on the shoulders of giants" (Newton 1676)[1].

[1]Isaac Newton in a letter to Robert Hooke dated February 5, 1676.

Often in the evolution of scientific thought, young minds have had the courage to develop thoughts, theories, and studies that others had for years simply mentioned in passing. In this way, the sum of so many small steps and the consolidation of theories brought researchers at the beginning of the 20th century to elaborate theories that marked a decisive turning point that departed from classical, mechanistic physics and the worldviews associated with it.

Considering the number and importance of the diverse theories that have transformed the study of the cosmos in these last hundred years, we have chosen in this work to focus on a few, representative theories and propositions of physical cosmology that naturally lend themselves to philosophical reflection, namely: the Big Bang theory, the inflationary paradigm, the apparent fine tuning of the universe and the proposition of a multiverse to explain it[2]. These choices allow us to examine how scientific progress in a specialized discipline — which represents for many a merely vague reminiscence of their high school education — can also represent an intellectual paradigm shift with existential import for how man understands himself, his destiny, and his role within the universe. Our analysis will thus consider not only the scientific data, but also possible philosophical implications of said findings. We hope that the fruits of our modest reflections will provoke curiosity, consideration, discussion, and a new gaze toward the stars full of wonder.

2. The universe: birth, evolution and understanding

In the 1990's an earthquake struck Los Angeles in California during the night and caused a blackout. People rushed to the streets and noticed an unusual white band in the night sky; many of them got so worried that they started calling the police: in complete obscurity for the very first time in their lives, they were

---

[2] "*A scientific theory is an explanation of some aspect of the natural world that can, in accordance with the scientific method, be repeatedly tested, using a predefined protocol of observations and experiments. Established scientific theories have withstood rigorous scrutiny and are a comprehensive form of scientific knowledge*" (Schafersman 1997). In this respect, only the Big Bang theory can enjoy this status, as will become apparent later on.

finally able to see our stellar neighborhood, the Milky Way (Sharkeyaug 2008).

Fewer and fewer people still live in direct contact with Nature: this intermediacy entails lack of knowledge and of interest. Gone are the days when mankind used to wonder about the cosmos and our place in it, giving birth to many myths in order to make sense of natural phenomena. Far from being unexplainable, observations lacked the framework and mindset of the scientific method, whose codification by Galileo has allowed mankind to scrutiny Nature without recurring to myths. Every scientific leap that followed since then has revolutionized our understanding and enlarged our horizons, providing continuous stimulus through emergence of new questions. Peering through the fog of superstition, even the universe as a whole started revealing itself.

Until almost a hundred years ago, the known universe amounted to our stellar island, the Milky Way galaxy, and was thought to be an immutable, rigid scene on which natural phenomena take place in a mechanistic way. Between 1905 and 1916 Albert Einstein devised completely new foundations for our understanding of the cosmos and its very nature: the so-called theories of Special and General Relativity. The universe was now a dynamic place, with a malleable scene resulting from the fusion of deformable space and mutable time. New phenomena were now possible, too extreme even for Einstein's genius, such as an evolving universe characterized by granularity and probability alongside continuity and determinism.

After Einstein published the first article on his theory of gravity and abandoned the hypothesis of an eternal static universe, in 1922 Alexander Friedmann sought to apply the equation of general relativity to the dynamic evolution of the universe. Friedmann based his study upon two hypotheses about the universe: when observed on a scale large enough to consider clusters of galaxies and not just the individual stars, the universe is the same, both in every direction in which it is observed (isotropy) and from whatever point in the cosmos one observes (homogeneity).

In 1929, however, the American astronomer Edwin Hubble observed that galaxies recede from us at a velocity proportional to their distance. Since it would be scientifically illogical and unmerited to sustain that they recede from the Earth as the center of the universe, Hubble came to hypothesize that every galaxy experiences a receding motion from the others and that, therefore, the universe was expanding. Tracing this expansion back in time, it is reasonable to hypothesize that its origin was a unique_region in which all the matter was contained, a so-called singularity for the extremeness of its mathematical and physical properties. This theoretical speculation was first elaborated by Georges Lemaître in 1927: corroborated by Hubble's observations, it would later go by the name of the Big Bang theory. George Gamow further developed the theory, understanding it should have left a faint afterglow radiation in the present universe (Gamow 1953).

Gamow's theory was met with scant response in the scientific community until 1965, when Arno Penzias and Robert Wilson discovered a background radio interference during their work as engineers for the Bell Laboratories. The radiation they found was uniform in every direction, both toward the Sun and the other stars as well as toward the empty cosmos: physicists recognized it was the fossil afterglow of the Big Bang, although without giving proper credit to Gamow. After this discovery, the Big Bang theory became the most credible cosmological model. However, there were still evidential incongruities.

The homogeneity and isotropy of the cosmic background radiation indicates an overall equilibrium across the universe, which is very surprising because even in regions at antipodes of the universe, the same properties are found: for example, density of matter, cosmic background temperature, and relative abundance of chemical elements. When tracing the Big Bang evolution of the cosmos backwards, one realizes that these regions have never been so close to each other to be causally connected and thus homogenize their properties: how then can they look so similar?

Moreover, the cosmic background radiation reveals that the universe has an overall flat geometry, which, according to Einstein's general relativity, corresponds to a very specific energy content of the universe itself: in order to maintain flatness during the entire evolution of the universe, this value should be fixed with very high precision at the moment of the Big Bang.

In 1981, the young physicist Alan Guth attempted to resolve the critical issues that had emerged in the standard Big Bang model. In his inflationary paradigm, Guth hypothesized an extremely fast and violent accelerated expansion of the universe, in which the distance between any two points of the universe would grow by a humongous factor. Thus, even two points that are at antipodes in the current universe could have been in contact at some moment in the past and would have had the chance to reach the equilibrium they show when observed today. By its very nature, Inflation would also stretch the three-dimensional space, thus smoothening any initial curvature wrinkle and explaining the measured overall flat geometry of the universe.
To understand better the aforementioned concepts, we can imagine an inflating balloon, whose surface is decorated with points that, instead of swelling ever more slowly, rather expand increasingly rapidly. To the extent that the distance between all of the different points grows, the balloon stretches its curvature, approximately flattening larger and larger regions as the process continues.
Thanks to its desirable features Inflation became a very popular framework among cosmologists. However, a more detailed explanation of how the general inflationary mechanism takes place in the physical reality is still escaping the grasp of scientists: while their models get more baroque, ever more precise data seem to indicate a simple universe. What conditions were present before cosmic inflation and in what manner could they influence the forecasts of the inflationary model? The question still awaits a response.

3. Fine tuning

The astronomer and astrologer Nicolaus Copernicus (1473-1543) is credited with overturning the idea that planet Earth stood motionless at the center of the universe: "Copernicus understands and shows that our Earth is not at the center of the dance of planets, but instead, that the sun is at the center. Our planet becomes just one of many" (Rovelli 2014). This great revolution, which seems to imply man's gradual exit from center stage and loss of his privileged role, represents an important passage from the medieval to modern age. After this conceptual revolution and throughout the subsequent years, with the birth of modern cosmology, the discovery of the evolution of the species, and impersonal demands of the modern scientific method, the role of man seems to take second place with respect to the immensity of a universe in continual expansion.

Only since the second half of the 20th century has man's position within the cosmos been emphasized with renewed vigor through the philosophical construct of the "anthropic principle." On the one hand, no one distrusts man's decentralized position within the physical universe anymore. On the other hand, man seems to occupy a decisively privileged position as a living being. It is both interesting and revealing that by coining the term "anthropic principle" one is assuming human life as the synonym of intelligent life. The fundamental principles of physics that dictate the structure and the laws of the universe, together with the delicate equilibrium of the evolutionary phases that brought about planetary and stellar formation, emerge as determinate and indispensable factors for the existence of life itself. In particular, the most reliable cosmological history currently available seems to suggest that a combination of innumerable "coincidences," each equally fundamental, permits our existence and allows our planet to be the only cradle that nurture intelligent beings. It seems as if the cosmos were finely tuned for our life: the structure of our universe, from the vastness of galaxies to the smallness of subatomic particles, is determined by fundamental constants that turn out to assume just the exact values needed in order to allow for the existence of a (human) life-permitting universe. A constant is "a fixed number which is everywhere and at every time the same throughout the history of the universe. These constants control the equations of physics (which describe the laws of nature). Therefore, the constants control the laws of physics in the universe." (Spitzer 2013). If but one of these constants were to suffer an infinitesimal variation of its precise value, then our life in the universe would not be possible.

There are more than 20 fundamental constants, which include: the Hubble constant (H = 69.32 ± 0.80 (km/s)/Mpc), the speed of light (c = 299,792.458 km/s), and the gravitational constant (Spitzer, 2013). A minutely small variation of the values of the fundamental constants would render our universe uninhabitable and without life. Spitzer makes the following example:

> "If the gravitational constant or the weak force constant had varied from their values by only one part in 1050 (0.00000000000000000 000000000000000000000000000000001) higher or lower, the universe would have either continuously exploded in its expansion, or contracted into a black hole (where the entire mass-energy of the universe would collapse into 10-33 cm). Either scenario would have been disastrous for the development of any life form. Let us try to find an adequate explanation of our "anthropic universe" (Spitzer 2013).

Some scientists have given this paradox an explanation that embraces a consonance with certain religious beliefs, thus offering a fundamentally finalistic interpretation of man's creation. This reading implies that the universe was planned to host life in accord with an "intelligent design" of a creator. As American physicist Leonard Susskind notes in his work The Cosmic Landscape: String Theory and the Illusion of Intelligent Design, "Our universe is an exceptional place that looks extremely accurate designed to make our existence possible, and this peculiarity is unlikely to be attributed to the case. These apparent coincidences are loudly explained by an explanation" (Susskind 2006).
Entering into the religious implications of such scientific research, the author continues, "A story that is very successful, not only with the general public but also with many scientists, is that the universe has been designed by a "superarchitect" driven by benevolent intentions. Supporters of intelligent design say that this is an absolutely scientific hypothesis that perfectly explains both the facts of cosmology and those of biology. The clever designer not only chose physics laws perfectly adapted to its purpose, but also guided the unlikely chain of biological evolution, from bacteria to Homo sapiens" (*ibidem*).
However, according to the same Susskind, the aforementioned explanation of the cosmological questions cannot be retained as intellectually satisfying, even if it may be emotively consoling: "Certain questions remain, such as those who have designed the

designer, through which mechanism intervenes to guide evolution, whether to pursue his or her purposes violates the laws of physics, and whether he himself is subject to the laws of quantum mechanics" (*ibidem*).

Hence, Susskind and others believe the anthropic principle can be justified by postulating "trillions upon trillions of 'other universes' (in a multiverse which is unobserved and likely unobservable)" (Spitzer 2016). According to the multiverse theory, there are innumerable universes parallel to our own. These universes could therefore possess among them all of the possible combinations of fundamental constants. With such a vast array of differing universes, the perfect harmony necessary for life is no longer improbable. Thus according to the multiverse hypothesis, some universes, including our own, are a sort of outlier that possesses the precise properties that permit life's appearance (Wolchover, Byrne 2014). The multiverse theory presents an explanation of the universe that *a priori* seems improbable and remains highly debated. Various scientists seek to demonstrate that our universe, among the other parallel universes that permit life, is in reality statistically typical (*ibidem*). Consequently, they work to show that the values of the fundamental constants of our anthropic universe have an elevated probability. Physicist Joseph Weinberg has applied the anthropic principle as an ex-post selection constraint by which our very existence forces the universe to possess the very specific characteristics that enable it, such as the value of the cosmological constant Lambda. The existence of intelligent life forms (observers) requires the presence of galaxies, which in turn can only form as aggregate of stars under determined values of the cosmological constant (*ibidem*). Other scientists, such as Donoghue and his colleagues, have used the process of anthropological selection to calculate the small mass of Higgs boson (*ibidem*): they have discovered a link between this value and the existence of carbon (a fundamental component for life), but only if this value is within a very narrow range (*ibidem*).

However, scientists have realized that theories based upon a process of anthropic selection are not sufficient to explain our universe in so far as they are limited in distinguishing the universes that have observers from those that do not (*ibidem*). The process of anthropic selection is based solely on the existence of observers and makes it highly improbable that a universe like ours should exist. Later, other theories were developed in order to compare galaxies among each other on the basis of the number of observers to then be able to demonstrate that our universe has an elevated probability of existence (*ibidem*). The galaxies that have the highest number of observers are those that have the highest probability of existing and thus of having life. Among such theories is that of Bousso, who sustains that a valid measure for estimating the number of observers is entropy, a quantity related to disorder, or degradation of energy (*ibidem*). In short: the higher the level of entropy, the higher the probability that there are observers. A universe with observers is complex and can permit life. Bousso goes further asking himself, cosmologically speaking, "Why do we happen to exist just now?" The researcher sustains that we are in a moment in which the probability of existing is high (*ibidem*).

Returning to the theory of fine-tuning, the fundamental constants must have a highly precise value for us to live in our anthropic universe. Consequently, it would seem, we can interpret the particular and "perfect" characteristics as a sign of the existence of an intelligent creator. However, the multiverse theory offers other responses that do not allow for the possibility of any sort of creative intervention. The theory that holds the coexistence of parallel universes in which the constants can have all possible combinations requires that there should be other universes similar to our own.

In conclusion, the aforementioned theories seek to mitigate man's singularity, thus liberating us from anthropocentric prejudices. The theory of fine-tuning suggests that our universe is special and discourages us from reducing our reflections to a discourse of probability. The multiverse theory seeks to liberate us through presenting our universe as merely one of many. However, by basing themselves upon the figure of the observer, these theories themselves risk falling into the dreaded anthropocentrism they purport to surpass.

4. Philosophical interpretations derived from cosmological data

While there is general agreement among scholars of contemporary cosmology regarding the fine-tuning of the universe as outlined in the above sections of this text, the challenge remains to interpret properly the scientific findings by themselves and in relation with other domains of human knowledge. Rashly drawing dubious theological conclusions from cosmological models, far from aiding the cause of faith, renders a profound disservice to the credibility of any institution willing to manipulate data for its own aggrandizement. That said, it would run contrary to man's natural aspiration to form a comprehensive understanding of the universe to impede him from drawing philosophical conclusions from the most reliable cosmological models available today. Following contemporary philosopher William Lane Craig[3], this section will explore and assess the merits of the three viable hermeneutics for understanding how technical advances in our vision of the cosmos might affect our anthropological conceptions (Craig 1990).

4.1 Physical Necessity

Perhaps the natural world is characterized by the physical constants and qualities it enjoys simply because these are the only combinations possible. As observers, we may be surprised that the universe's attributes fall within such a narrow range of possibilities, when in effect, there are no other possible universes. Physicists Barrow and Tipler try to explain this ex-post observation by means of the Weak Anthropic Principle making the following example:

> Even though the probability of a given Briton being monarch is about 10-8, someone must be. Only if there is a monarch is it possible for the monarch to calculate the improbability of her particular existence. Similarly, only if an intelligent species does evolve is it possible for its members to ask how probable it is for an intelligent species to evolve. Both are examples of WAP self-selection in action (Barrow, Tipler 1986)

[3] The present section of the text is indebted to Craig for the sources of contemporary cosmology he has gathered and for his philosophical analysis.

Therefore, the world simply is as it is by physical necessity and our efforts for some overarching explanation are a pretense doomed to failure. Perhaps we are guilty of a certain anthropocentrism that unwarrantedly assumes that some mysterious force cares for our existence. Diverse and various religious myths betray the human urge to position man's history inside the plan of supernatural powers concerned with humanity's destiny. Nonetheless, the cold precision of physical research could purify such antiquated religious notions to give man a more realistic notion of his place in the universe as mere random aggregate of atoms. In this scenario, we can reduce man's biology to his chemical components, which in turn are determined according to the impartial laws of physics. While the shift from anthropocentric creation myths will entail a bit of psychological dissonance for those accustomed to finding comfort in the presence of divine in their lives, the frank admittance that our universe simply is the way it is could eventually liberate man from puerile illusions. Belief in Santa Claus and the Easter Bunny is well enough for an imaginative child, however, we could only pity a grown man whose life hinges upon gifts left to him by a bearded visitor from the north pole.

Proponents of the physical necessity hypothesis discourage facile anthropocentric explanations for complicated scientific data. The assessment of the universe's composition should never depend upon soothing stories or wishes. Rather, we should confront reality squarely, even when it fails to conform to our preferences or sense of importance. Nonetheless, it seems unsatisfying to explain away cosmic fine-tuning via appeals to a universe that is the way it is because it cannot be otherwise.

In fact, the constants and qualities of the cosmos described in previous sections are independent of the known laws of physics and could thus exist in very different combinations from those required by our intelligent-life-permitting universe. Although there may not be any grounds for affirming the actual existence of other universes with distinct constants, imagining their possibility entails no logical contradiction. Speculations regarding possible alternative universes may seem useless apart from possible science fiction scripts. However, while the consideration of possible alternative universes may not aid our understanding of our current universe's makeup, it can provide some helpful indirect clues regarding the implications of the fine-tuning under consideration. Cosmologists, in fact, can envision a host of other universes with varying constants and qualities that would be able to function without the precise conditions necessary for our instance of intelligent life (Craig 1990). Therefore, the possibility of alternative realities confirms that physical necessity is not enough to account for the particular combination of constants we find in the world around us.

4.2 Pure Chance

If the possibility of alternative universes disqualifies physical necessity as an adequate hermeneutic for cosmological fine-tuning, then perhaps it is more fitting to attribute the delicate balance of life-permitting conditions to chance. The pure chance explanation of cosmic fine-tuning amounts to admit that, given the array of possible universes inhospitable to life, it is rather fortunate that we should find ourselves in a world that sustains intelligent beings like ourselves. Assuming the existence of intelligent life is preferable to its non-existence, we can celebrate our luck in winning the cosmic lottery. While this might inspire in man a sense of deep gratitude, there is not necessarily anyone or anything toward which to direct our thankful sentiments.

Moreover, subjective emotions of surprise are rationally ungrounded upon further reflection. There is little sense considering the alternative of existing in the many possible inhospitable universes, since their hostile physical conditions would prevent our very existence, let alone our capacity to form cosmological thinking. We simply happen to be in the only type of universe that allows ourselves as observers capable of considering other cosmological possibilities.

Other cosmologists are willing to admit that simply reminding observers that they exist in an observer-friendly environment is not enough to explain away the unlikelihood that such situation should exist rather than others. These other situations would therefore constitute some form of the world-ensemble theory in which the aforementioned alternative universes of diverse constants are not just theoretical possibilities, but real ones. Each possibility represents a path that can be taken by the universe: in reality, each path would be taken by one branch of the universe and become a new independent reality. All these realities would exist simultaneously and in parallel to the universe of our current experience (Everett 1957). In such a vision, the vast array of possible universes has actually been realized. While most of the other existing universes are indeed unable to sustain intelligent observers like ourselves, we are clearly in a version of the realized possibilities that permits our emergence and survival. If there is an infinite or nearly infinite number of universes, then it should not be at all surprising that at least one of them happens to possess those highly precise and intricately aligned constants and qualities needed for intelligent life. Thus, it is unfair to describe our world as somehow privileged in comparison with other possible alternatives, since those other possibilities have and are being realized as we speak. We need not appeal to supernatural forces to explain any preference given to the delicately balanced life-permitting constants in favor of the wide range of other possible configurations, since in the world-ensemble model, our life-friendly universe enjoys no existential superiority over and above the plethora of equally existent universes bereft of intelligent life.

It is worth noting that the world-ensemble model is based on the assumption that multiple universes are inflationary. The cosmic inflation mechanism proposed by Alan Guth has it that immediately after the Big Bang, a sort of repulsive gravity made our universe expand at an enormous pace. Only afterwards did it reach a steadier pace of expansion, which it currently retains. Thanks to the gravitational repulsion and the energy conserved during this initial phase of rapid inflation, parts of what is now our observable universe would evolve into other bubble universes – creating an eternally expanding multiverse with various sets of physical laws (Guth 2013). Guth maintains that the conservation of energy typical for inflation makes the process endless, however, all expanding inflationary universes unavoidably have a beginning at some point in the past, as demonstrated in the Borde-Vilenkin-Guth proof:

> We made no assumptions about the material content of the universe. We did not even assume that gravity is described by Einstein's equations. So, if Einstein's gravity requires some modification, our conclusion will still hold. The only assumption that we made was that the expansion rate of the universe never gets below some nonzero value, no matter how small. This assumption should certainly be satisfied in the inflating false vacuum. The conclusion is that past-eternal inflation without a beginning is impossible (Vilenkin 2006).

As long as the average expansion rate in an inflationary multiverse is greater than zero, the number of possible parallel universes is finite. Even if this fact does not prove the world-ensemble model fallacious, it decidedly limits its scope. What additionally undermines the multiverse theory is the Ockham's razor rule – the principle according to which the best explanation of a problem is the one entailing the least number of assumptions (Spitzer 2016). As Paul Davies argues, world-ensemble models require a nearly unlimited number of possible universes for our single life-permitting universe to exist. It is impossible either to confirm or to reject such an idea; however, there is no scientific evidence which could exclude the plausibility of the Divine creator (*ibidem*).

Given that lack of probative support for the world-ensemble theory, the supposition that the subtle balance of life-permitting constants and forces in our universe has been a work of pure chance appears highly debatable. In fact, on one side the range of values within which cosmological constants could possibly fall at the Big Bang is enormous; on the other, the precise values the constants should take to allow for the existence of intelligent life are extremely narrow-range, hence many scientists doubt that the intricate network of life-permitting factors occurred in our universe by mere coincidence.

By analyzing the so-called "anthropic coincidences", one may discover that even unimaginably tiny variations in the values of particular constants would suffice to change the structure of the universe in a dramatic way. If the gravitational force were a bit weaker and electromagnetism only slightly stronger, the sole stars would be red dwarfs; if, on the other hand, gravitation were stronger and the electromagnetic interaction weaker, only blue giants could exist: both types of stars produce insufficient amount of light and heat to sustain life-friendly planets. Likewise, an unbelievably small difference in the values of either gravity or the weak force would make our universe collapse at the very beginning of its existence or result in its explosion while expanding too rapidly for any galaxies to form. Similarly, if there were only a 2% change in the value of the strong nuclear force constant, the universe would either lack hydrogen, and the life-essential water, or would be bereft of the carbon equally necessary for the development of various life forms. Keeping this in mind, a number of physicists and philosophers emphasize the fallacious character of the pure chance model and maintain that a possible clarification regarding the origins of the universe should be sought somewhere else. One of the alternatives they propose is a design model, which contends that a transcendent creator of enormous power is ultimately responsible for the existence of intelligent life and the universe as such (Spitzer 2016).

The world-ensemble model provides an intriguing vision of the cosmos, which is more rich and varied than previously imagined and should not be dismissed, even if only for expanding our vision of reality. After all, quantum mechanics introduced us to realities once unimaginable for the limits of our Newtonian models. Nonetheless, while the world-ensemble model laudably avoids unthinking invocations of the "god of the gaps" to resolve explanatory difficulties eventually soluble through the development of more thorough cosmological models, it suffers the weakness of being a highly speculative supposition lacking in evidence. In principle, the branching universes are unobservable and exercise no influence upon our current cosmos. Were they otherwise, they would be part of our universe and subject to the fine-tuned conditions the model seeks to explain away. Yet, such radical separation from our known world prevents the very observational and experimental testing upon which modern science depends. In their effort to avoid pre-scientific myths, multiverse theories ironically tend to degrade into a-scientific, metaphysical, and unfalsifiable presuppositions.

4.3 Design

Since the two warring hermeneutics for the cosmic fine-tuning of the universe, namely necessity and chance, have been found wanting, it remains to be examined whether the residual viable option of design enjoys sufficient evidence as the most adequate interpretation for the findings of contemporary cosmology. According to the design conjecture, some form of transcendent intelligence is reasonable for ensuring the delicate balance of initial conditions of the universe that permit intelligent life. To visually grasp the credibility of the design supposition, we could imagine a complex studio panel with, say, twenty dials. If each dial were found at exactly the right setting amid a wide range of possibilities and if all the dials were found to be simultaneously coordinated among each other exactly to the right setting conducive to perfect sound, it would be most reasonable to assume that an intelligent technician or producer prepared the panel so that the intentionally willed harmonious music might resound in pristine beauty. It would be unreasonable to assume that random effects like draft just so happened to guide each dial exactly to that perfect arrangement among innumerable possible cacophonous combinations. This situation amounts to the seemingly inexplicable fine-tuning of the physical constants.

The design hypothesis explored in the present work is distinct from the "god of the gaps" that some proponents of the Intelligent Design movement invoke to account for the organic complexity currently inexplicable in light of current biological knowledge (Craig 1988).

The design hypothesis for cosmic fine-tuning makes an explanatory inference from established features present from the first phases of the universe. It therefore does not compensate for present ignorance that future scientific advances will likely explain away naturalistically.

In affirming the probable existence of an intelligence behind cosmic design, we should not be too confident of our knowledge of said intelligence being. As the guiding force behind the initial conditions of the universe that continue to guide nature, the intelligence seems to transcend the spatio-temporal limitations of our cosmos. In that case, it would be an eternal spirit of at least exceedingly great power. It would be an unwarranted jump to claim that the design hermeneutic of cosmic fine-tuning is equivalent to the Divine found in major religious traditions. Nonetheless, the existence of such a transcendent intelligence remains distinct, though compatible, with various theistic traditions. Consequently, the design hermeneutic poses a challenge to reigning naturalistic or scientist worldviews that exclude such beings *a priori*.

While theistic and atheistic thinkers are held to the same canons of scientific rigor in assessing the material universe, the metaphysical conclusions they may draw differ and should be evaluated according to the standards of right philosophical reasoning. Science and Philosophy, even though methodologically different fields, are by no means conflicting areas of thought. To illustrate this idea, it suffices to evoke Albert Einstein. Even though an undeniable giant of science, he remained open to the plausibility of a transcendent being standing behind the creation of the universe: "You may call me an agnostic, but I do not share the crusading spirit of the professional atheist… I prefer an attitude of humility corresponding to the weakness of our intellectual understanding of nature and of our own being".

We therefore face a surprising reverse in the standard scientist narrative that charts a constant process of secularization parallel to the developments of modern science. The dominant secular narrative contends that the more we know about the natural world, the less need we have to invoke God as an explanatory hypothesis (Barr 2003). The naturalistic account argues that scientific advances have continually undermined claims of man's unique dignity as a superior being in the cosmos. Copernicus, in positioning the Sun rather than the Earth at the center of the universe, seems to have removed man from his privileged center. Newton, Galileo, and the other heroes of the Scientific Revolution overcame primitive Aristotelean notions of purpose in nature with their more rigorously accurate mechanistic models. Darwin is said to have exposed and definitely refuted man's pretensions to be anything more than a complex beast of nature. The rational man of the scientific age naturally cringes before his own insignificance, but is called to embrace his irrelevance in the cosmic flux of the universe. Though the naturalistic worldview may rob man of the crutch of religious sentiment, he is at least free to limp toward the abyss of death, free from the mental and moral impositions of organized religion.

While no cosmological model can or should ever be used to demonstrate religious tenets, our cursory evaluation of leading philosophical interpretations of contemporary cosmology indicates that naturalism is by no means the indisputable worldview for today's reasonable person. Just as religious believers can sometimes lamentably fall into fundamentalist views that reject the findings of science in a misguided quest to protect religion, so too, some convinced naturalists can manipulate scientific data to fit their prior philosophical commitments. With candor worth citing at length, Harvard University Biology professor Richard Lewontin explains just how profoundly philosophical presuppositions can condition leading scientists:

> Our willingness to accept scientific claims that are against common sense is the key to an understanding of the real struggle between science and the supernatural. We take the side of science *in spite* of the patent absurdity of some of its constructs, *in spite* of its failure to fulfill many of its extravagant promises of health and life, *in spite* of the tolerance of the scientific community for unsubstantiated just-so stories, because we have a prior commitment, a commitment to materialism. It is not that the methods and institutions of science somehow compel us to accept a material explanation of the phenomenal world, but, on the contrary, that we are forced by our *a priori* adherence to material causes to create an apparatus of investigation and a set of concepts that produce material explanations, no matter how counter-intuitive, no matter how mystifying to the uninitiated. Moreover, that materialism is absolute, for we cannot allow a Divine Foot in the door (Lewontin 1997).

The activity of a transcendent designer of the universe seems to offer a philosophically superior interpretation of contemporary cosmology in comparison to the physical necessity or pure chance alternative hermeneutics. There therefore seems to be rational warrant in adopting a worldview open to the transcendent. In contrast, contemporary cosmology does not seem to lend support to the naturalistic narrative of man's dethronement through science's advances and its reductive vision of man as a mere randomly arranged conglomerate of physically determined molecules.

The design interpretation of the cosmic fine-tuning of the universe is far from the only philosophical approach to God. Some theistic philosophers fear that the design argument places too much weight upon changing cosmological models that could be drastically revised in light of future discoveries (Beckwith 2017). They would avoid probabilistic arguments sympathetic to a transcendent intelligent designer in favor of metaphysically demonstrative arguments based upon the ontological contingency of the universe that demands the ongoing conservation of an ontologically self-sufficient Absolute Cause. An assessment of the relative merits of such metaphysical arguments goes beyond the scope of our modest reflection upon the most reasonable philosophical interpretation of cosmological fine-tuning. Thinkers like Spitzer, affirm both the design argument based upon cosmic fine-tuning as well as metaphysical arguments as two complementary paths to understand better the transcendent reality ultimately responsible for creating the universe we inhabit (Spitzer2010).

5. Conclusions

The various cosmological theories of the past and of today reveal how science has always sought the response to the question of the origin of the universe. Scientific research, based upon observation, pushes man toward a study that provides an ever more profound knowledge of the marvels that surround him.

The questions that are raised resurface in every age and led us to ask: Who has designed the universe? Moreover, if there is indeed a creator, is there another responsible for the "super-architect" of good will? Is it not perhaps easier to assume that the universe has existed forever without any need for creation at a given moment? If this universe did begin at a determined moment, what then existed before said starting point? Was there really a Big Bang? With all of our recent scientific advances, why have we only succeeded in explaining five percent of the entire universe? How did cellular life really begin?

The multiplicity of proposed cosmological theories and interpretations renders it difficult for even the educated person to arrive at a definitive grasp of the world's origins and his place in the universe. A number of scientists have argued that their research excludes the credibility of a divinity. They present liberation from antiquated notions of a God figure as the crowning achievement of scientific progress. For instance, the renowned physicist and popular author Stephen Hawking declared that "the fundamental questions of life have a response and we are near the truth about the Great Design: there is no Designer!" (Hawking 1993). Appealing to quantum mechanics, the same Hawking together with physicist James Hartle come to hypothesize the existence of a multitude of universes coexistent with our own, each with its own physical laws, fundamental constants, and spatio-temporal dimensions. They are thought to have been born spontaneously through so-called quantum fluctuations, infinitesimal hiccups of nature at the microscopic level. The physicist Mlodinow, not without controversy, contends that science is capable of demonstrating that the universe may have been created from nothing through the mere laws of physics. It is therefore not necessary to "invoke God to light the blue touch paper and set the universe going." Mlodinow instead argues "spontaneous creation is the reason there is something rather than nothing, why the universe exists, why we exist." (Hawking - Mlodinow 2010). Similarly, the Italian astrophysicist Margherita Hack explains the origin of the universe from empty space. She writes as follows:

An idea of the universe that I find particularly acceptable is that there was a vacuum, but not nothing; the vacuum of the physicists in which particles and antiparticles are continually annihilated thus freeing energy that recreates other particles and antiparticles in an infinite space and time; an infinite space and time that does not exclude the possibility that new regions are continually formed like the one in which we live and call the universe, but is not the only one. (Hack 2009).

Hack recognizes that such an explanation may not convince many, but she insists that such a position is preferable to an irrational appeal to a mythological celestial father figure (*ibidem*).
Today more than ever the celebrated phrase of the father of microbiology rings true: "A little science estranges men from God, but much science leads them back to Him."[4].
Our own modest study has sought to provide an alternative narrative to popular materialist accounts. We have shown that philosophical appeals to necessity and chance are inadequate explanatory hypotheses. While the scientific study of the material world can by its very nature never demonstrate a metaphysical reality, our philosophical reflection upon the latest cosmological data reveals that inference to design provide a credible explanation for the fine-tuned features of our cosmos.

[4] Louis Pasteur considered the founder of modern microbiology, chemist and also interested in physics. He lives a fervent Christian and observes the great religiosity of the Bretons, expressing this desire: "I have the faith of a Breton peasant and for the moment I die I hope to have the faith of the wife of the Breton peasant."

References

AGULLO I. – ASHTEKAR A. – NELSON W., «Quantum Gravity Extension of the Inflationary Scenario», in *Physical Review Letters*, 2012, 109.

BARR S., *Modern Physics and Ancient Faith*, University of Notre Dame Press,2003, 19-21.

BARROW J.D. – TIPLER F.J., *The Anthropic Cosmological Principle*, Clarendon Press, Oxford, 566.

BECKWITH F.J., *God Is Not a Scientific Hypothesis*, 2017 *www.thecatholicthing.org*

CRAIG W.L., «Barrow and Tipler on the Anthropic Principle vs. Divine Design», in *British Journal for the Philosophy of Science* 38, 1988, 389.

CRAIG W.L., «The Teleological Argument and the Anthropic Principle», in *The Logic of Rational Theism: Exploratory Essays*, Edited by Wm. L. Craig and M. McLeod. Problems in Contemporary Philosophy 24. Lewiston, N.Y.: Edwin Mellen, 1990, 127-153;

DAVOUDIASL H. – HOOPER D. – MCDERMOTT S.D., "Inflatable Dark Matter", *in Physical Review Letters,* 2016, 116.

EVERETT H., «Relative State' Formulation of Quantum Mechanics», in *Reviews of Modern Physics* 9, 1957, 454-462.

GUTH A., *LECTURE 1: INFLATIONARY COSMOLOGY: IS OUR UNIVERSE PART OF A MULTIVERSE?*, PART I, 2013, www.ocw.mit.edu

HACK M., «Siamo figli delle Stelle», in *Messaggero*, 8 gennaio 2009.

HAWKING S.W., *Buchi neri e universi neonati, Riflessioni sull'origine e il futuro del cosmo,* Rizzoli, Milano, 1993.

HAWKING S.W. – MLODINOW L., *Il grande disegno. Perché non serve Dio per spiegare l'universo*, Mondadori, Milano, 2010.

LEWONTIN R., *Billions and billions of demons (review of The Demon-Haunted World: Science as a Candle in the Dark by Carl Sagan, 1997)*, The New York Review of Books. 1997, 31.

NBC, *Letters From Einstein About God (and Toys) Sell for $420,625*, www.nbcnews.com

NEW YORK TIMES, SHARKEYAUG J., *Helping the Stars Take Back the Night*, 31/08/2008, www.nytimes.com

ROVELLI C., *Sette brevi lezioni di fisica*, Adelphi Edizioni, Milano, 35, 2014.

SCHAFERSMAN S.D., *An Introduction to Science Scientific Thinking and the Scientific Method*, Miami University, 1997

SPITZER R.J., *New Proofs for the Existence of God: Contributions of Contemporary Physics and Philosophy*, Wm. B. Eerdmans Publishing, Grand Rapids, 2010, 23.

SPITZER R.J., *Fromt Nothing to Cosmos: God and Science, A comprehensive Look at the Evidence for God*, Magis Center, Ph.D, 2013, 56, 327. www.magiscenter.com;

SPITZER R.J., *Evidence of God from Contemporary Science & Philosophy*, Magis Center, Ph.D., in: www.magiscenter.com, 2016, 23.

SUSSKIND L., *The Cosmic Landscape: String Theory and the Illusion of Intelligent Design*, Little, Brown and Company, NY, 2006.

VILENKIN A., *Many Worlds in One: The Search for Other Universes*, Hill and Wang, New York, 2006.

WOLCHOVER N. – BYRNE P., *In a Multiverse, What are the Odds?*, Many Worlds in One: The Search for Other Universes. By Alexander *Vilenkin*. New York: Hill and Wang, *2006*. *Quanta Mgazine, Series: Infinity and Beyond: the Ultimate Test*, www.quantamagazine.org.

Afterword

"Beyond the limits. Genetic Neuroscience and the Technological Revolution" already the title of this publication suggests, that the work of the "Sinderesi School" and the discorse of social ethical issues has entered into new spheres and has opened new areas of discussion. For six years the Konrad-Adenauer-Stiftung in Italy has been encouraging the work of young people under the great supervision of Mons. Samuele Sangalli and with the support of qualified and engaged professors and experts. Together the students have promoted the discourse and the dialogue on social and ethical issues and have grown within the group and as a group. I want to thank the students and recognize the academic experts for their important and tireless work. A great appreciation also goes to the "Pontificial Gregorian University" and the "Alberto Hurtado-Centre" for the trustful cooperation over the last years.
The Konrad-Adenauer-Stiftung (KaS) runs programs in more than 100 countries on four continents, and has about 80 offices around the world. We consider public debate as a key element for the democratic decision-making process. The exchange with young people is important for us and need to be encouraged. This publication wants to offer new perspectives on different topics and spark new ideas. The Success of the "Sinderesi – project" is also due to the fact that the in recent years it always strived for higher goals. We hope that the students will carry on their work in the future and with great motivation and enthusiasm!

Caroline Kanter
Director of the Konrad-Adenauer-Stiftung in Italy

THE PATH OF SINDERESI SCHOOL

BEYOND THE LIMITS

Consequences of technological revolution in society

Coordinator of the School: Mons. Samuele Sangalli
Assistant: Dr. Antonella Piccinin

Part One: lectures (from October to December):
PATHS OF THINKING

Saturday 15 October 2016 (4.00 - 7.00 pm)
PHILOSOPHY – *The Conversion of the ego as priority of the Techno-Age.*
Mons. Samuele Sangalli,
Coordinator of Sinderesi School – Pontifical Gregorian University (Rome)

Saturday 19 November 2016 (from 4.00 pm to 7.00 pm)
LAW – *Genetics and Intellectual property*
Dr. Roman Cholij,
Research Associate Von Hugel Institute – Cambridge University (UK)

Saturday 17 December 2016 (from 4.00 pm to 7.00 pm)
SCIENCE – *The Cyborg and Cyborgization*
Prof. Paolo Benanti, TOR
Faculty of Theology - Pontifical Gregorian University (Rome)

Part Two: workshops (from January to May):

PATHS OF INVESTIGATION

Saturday 21 January 2017 (from 4.00 pm to 7.00 pm)
SPIRITUALITY – *Neurosciences and the search for "the Person"*
Case Study: *Criminal Law and Human Structure*
Under the supervision of Prof. Franco Imoda SJ,
Pontifical Gregorian University (Rome)

Saturday 18 February 2017 (from 4.00 pm to 7.00 pm)
POLITICS – *Technology and democratic procedures*
Case Study: *How technology shapes democracy in the social networks and "Post-Truth" era.*
Under the supervision of Prof. Giovanni Moro,
Università di Roma Tre (Roma) - Pontifical Gregorian University (Rome)

Saturday 18 March 2017 (from 4.00 pm to 7.00 pm)
BIOLOGY – *Towards an alternative Bioethics*
Case Study: *When does human life begin?*
Under the supervision of Prof. Ramon Lucas Lucas LC,
Pontifical Gregorian University (Rome)

Saturday 22 April 2017 (from 4.00 pm to 7.00 pm)
ECONOMICS – *Changes in the Financial System*
Case Study: *Impact Investing*
Under the supervision of Prof. Fernando De La Iglesia SJ,
Pontifical Gregorian University (Rome)

Saturday 20 May 2017 (from 4.00 pm to 7.00 pm)
COSMOLOGY – *After quantum physics: shaping a new cosmology*
Case Study: *Cosmic inflation and fine tuning*
Under the supervision of Dr. Umberto Cannella,
Swiss Space Center, Polythecnic Federal Institute (Lausanne)

Finito di stampare nel mese di ottobre 2017
presso Printbee.it - Noventa Padovana (PD)